In search of a home

In search of a home
Rental and shared housing
in Latin America

ALAN GILBERT
University College London

IN ASSOCIATION WITH
Oscar Olinto Camacho
René Coulomb
Andrés Necochea

The University of Arizona Press
Tucson

363.58
G464i

© Alan Gilbert 1993

First published in 1993 by UCL Press

UCL Press Limited
University College London
Gower Street
London WC1E 6BT

The name of University College London (UCL) is a registered
trade mark used by UCL Press with the consent of the owner.

First published in the United States of America in 1993 by
The University of Arizona Press
1230 N. Park Avenue, Suite 102
Tucson, Arizona 85719

ISBN 0-8165-1388-0 ⟨P

Library of Congress Cataloging-in-Publication data have been requested.

Typeset in Palacio (Palatino).
Printed and bound in Great Britain

Contents

List of tables

List of figures

Glossary of acronyms and names of institutions

AD	Democratic Action Party (Venezuela)
AURIS	Urban Action and Social Integration Agency (Mexico)
Banco Obrero	Workers' Bank (Venezuela)
BCV	Central Bank of Venezuela
Cámara Inmobiliaria	Chamber of Real Estate (Venezuela)
CBD	Central Business District
CENVI	Centre for Housing and Urban Studies (Mexico)
CEU	Centre for Urban Studies (Caracas)
CNOP	National Confederation of Popular Organizations (Mexico)
CONASUPO	National Company for Popular Subsistence (Mexico)
COPEI	Christian Democratic Party (Venezuela)
CORETT	Commission for the Regularization of Land Tenure (Mexico)
CORVI	Housing Corporation (Chile)
CTM	Mexican Workers' Confederation
FOGA	Fund for the Guarantee and Support of Housing Credit (Mexico)
FONHAPO	Trust Fund for Popular Housing
FOVI	Fund for Banking Operations and Discounts to Housing
FOVIMI	Housing Fund for the Military
FOVISSSTE	Housing Fund for State Workers
FUNDACOMUN	Foundation for the Development of the Community and Municipal Development (Venezuela)
INAVI	National Housing Institute (Venezuela)
INDECO	National Institute for Community Development and Housing (Mexico)

INFONAVIT	National Institute of the Fund for Workers' Housing
MINVU	Ministry of Housing and Urbanism (Chile)
OCEI	Central Office of Statistics and Information (Venezuela)
OMPU	Metropolitan Office of Urban Planning (Caracas)
PAN	National Action Party (Mexico)
PFV	Housing finance Programme (Mexico)
PRD	Democratic Revolutionary Party (Mexico)
PRI	Institutional Revolutionary Party (Mexico)
SERVIU	Regional Housing and Urbanization Service (Chile)
SINAP	National Savings and Loans System (Chile)
SNAP	National System of Savings and Loans (Venezuela)
UF	Index-linked finance unit introduced by Pinochet government
UNECLAC	United Nations Economic Commission for Latin America and the Caribbean
Unidad Popular	Popular Unity Alliance, 1970–73 (Chile)

Glossary of Spanish terms

allegado	Chilean term for those sharing a house with a resident family
barrio	Venezuelan term for an area of self-help settlement
bolívar	Venezuelan currency
callampa	"mushroom" settlement; term for invasions sometimes used in Chile
campamento	invasion settlement in Chile, usually applied to those formed by parties of the left
casa de inquilinato	rental tenement in Caracas
caseta sanitaria	sanitary unit (Chile)
comuneros	owners of communal land in Mexico
colonia	Mexican term for a residential neighbourhood
conventillo	rental tenement in Santiago
delegación	administrative area within the Federal District of Mexico
ejidatario	member of an *ejido*
ejido	community land created by 1917 Agrarian Reform Law in Mexico; also some public land in Venezuela
escudo	former Chilean currency
mediagua	provisional prefabricated housing unit provided under the Chilean Church's "Home of Christ" programme
Operación Sitio	Operation Site; programme launched by President Frei of Chile between 1964 and 1970
peso	Chilean and Mexican currencies
rancho	Venezuelan term for a shack
toma	Chilean term for a land invasion
vecindad	mainly Mexican term for a rental tenement
zócalo	central square, e.g. in Mexico City
23 de enero	huge high-rise apartment complex in Caracas

Acknowledgements

The research that underpins this book was funded by the International Development Research Centre of Canada. That institution proved both a generous donor and one that was always understanding about the delays and prevarications caused by academics, particularly those involved in a research project involving teams from four different countries. My only regret is that they have recently come under severe financial pressure and have been obliged to curtail the admirable urban development progr- amme. I would especially like to thank François Belisle, who initially encouraged this project, and Yvonne Riano and Luc Mougeot who later helped co-ordinate it when François was elevated to head the urban development programme.

 I should also like to thank my colleagues in the research project René Coulomb, Oscar Olinto Camacho and Andrés Necochea for their hard work and patience. Their efforts gathered the survey data used in this book, and their knowledge of their home cities was critical in helping me to understand better what was going on in three large and very complex urban areas. In this sense, this is very much a joint text. It is, however, only part of the overall research output, two of the teams having already published their results in their own countries (Camacho & Terán 1991, Coulomb & Sánchez 1991).

 Within Britain, I should like to thank University College London for its institutional support, and particularly Tim Aspen, who drew the maps, and Claudette John, who helped in various ways with the accounts and the administration.

 Finally, thanks are due to my family who have never entirely adjusted either to my absences abroad or the time I spend writing at home. Occasionally, they have even helped the writing process by keeping a little quieter in the house than is their normal practice.

LONDON
FEBRUARY 1992

CHAPTER ONE

Introduction

For many years the housing and planning literature has been fascinated by the phenomenon of self-help settlement. Writers have debated its merits and weaknesses as a means by which the poor can satisfy their basic shelter needs. In studying the self-help process, most of the literature has assumed that every family both wishes and will be able to become an owner–occupier. Given this focus, the large numbers of tenants and sharers living in Latin American cities have become invisible. Since a similar assumption has been inherent in the policies of most national governments, rental housing has seldom figured in recent housing programmes. While many governments introduced rent controls during the 1940s, little new legislation has been approved since. This is surprising given the apparent crisis affecting the rental housing sector. For many years, landlords have complained that investing in rental housing is a poor business, but governments have ignored their pleas. Tenants have demanded that landlords be required to improve the quality of their rental accommodation and the authorities have turned a deaf ear.

Rental housing has been ignored for three main reasons. First, there has been a dramatic shift since 1950 towards owner-occupation. Renting, the typical tenure of most urban Latin Americans before 1940, has been in rapid decline. Most governments have believed that every family would one day own their own home. They have concentrated on building finished houses and apartments or giving incentives to the private sector to build for lower-income groups; only comparatively recently have sites-and-services and squatter-upgrading programmes become more common. In the process, tenants and landlords have been ignored. Governments have let themselves believe in the myth of universal owner-occupation.

Secondly, most governments approved of home ownership because it was seen to be a stabilizing force in society. It was an important way by which the urban poor, migrants and natives alike, could establish a stake in the urban order. It seemed as if most families wanted to own their own home and governments were only too keen to encourage this aspiration.

1

People not only wanted their own homes, but owner-occupation was a useful political tranquillizer.

Thirdly, landlords and tenants are seldom natural allies. And, since the rental housing market rarely works well under conditions of poverty, most governments have, at one time or another, been forced to intercede over some kind of rental dispute. Unfortunately, their interventions have rarely proved very successful. When they introduced rent controls, the effect was neither efficient nor equitable. When they became landlords themselves, they often drew the wrath of the tenants onto their own heads. As a result, they tended to withdraw from rental housing; few if any Latin American governments now let public housing.

Only during the 1980s was there a slight shift in public attitudes. While we should not exaggerate the change, several Latin American governments began to express concern about the rental housing market. They became more attentive because they were worried that the so-called "austerity riots" might easily spill over into a rental housing sector suffering the effects of rapid price inflation (Walton 1989). Most importantly, however, they were forced to listen to the complaints of those who aspired to ownership but who could no longer afford it. Since 1980, many young middle-class families have had difficulty buying a home. For the poor, opportunities for self-help became more expensive and problematic. In many of the larger Latin American cities there are clear signs of a limit to universal home ownership, even of the self-help kind (Gilbert & Varley 1991).

There seem to be three major barriers to growing owner-occupation. First, whereas it used to be possible for the poor to invade land, that option has generally become less feasible. The authorities are attempting to control peripheral land. They have become aware of the problems involved in servicing low-density urban sprawl and have also been alerted to the environmental problems that uncontrolled suburban expansion can produce. The corollary of stronger controls over land invasion, however, is that land is becoming commercialized and the value of that land is rising (Angel 1983, Baróss 1983, Durand-Lasserve 1986, Trivelli 1986, Doebele 1987, Payne 1989, Baróss and van der Linden 1990).[1] Under such circumstances land for self-help housing has become less accessible. Secondly, as cities become larger so the problems of living in a peripheral settlement are likely to increase. For example, the cost of travelling from home to work, or of moving about the city generally, is rising both in terms of both time and financial outlay. With Mexico City accommodating

---

1 In fact, there is some evidence that the recession has cut land values in some Latin American cities (CED 1990, Jones 1991, Ward et al. 1991). Similarly, there is some evidence that real housing prices have fallen in certain places (Gilbert 1989). However, incomes have probably fallen more than land values or house prices. For a more extended discussion of this issue see Gilbert (1992).

around 15 million people and São Paulo some 12 millions, the difficulties involved in travelling around the city can be imagined. Clearly, the balance of advantage is shifting from peripheral locations to more central and consolidated areas. As a result, families will be less eager to live in the distant periphery. Thirdly, incomes in most Latin American cities have fallen dramatically over the past decade, a result of the debt crisis, the economic recession and the related austerity programmes (Lustig 1989, Portes 1989, World Bank 1990, González & Escobar 1991, Klak 1992). The fall in incomes has raised the real cost of self-help building because buying essentials such as food and clothing has taken up more of the family budget and because the costs of building materials, services and land have become more expensive.

The impact of the current recession on the housing situation varies from place to place and we will not know the precise effects until the results of the current round of census figures appear over the next few years. Broadly, however, there seem to be three likely outcomes (Gilbert 1989). The first is that the shift towards owner-occupation will continue but that the living conditions of owner households will deteriorate. There will be less money available for the middle class to buy decent homes and they will trade downwards. In turn, this will force more lower-income groups into self-help ownership, to join the millions already striving, with diminishing resources, to consolidate their homes. In addition, since the authorities have been paring their budgets, they have been unable to invest so much in infrastructure provision. During the 1990s, fewer self-help settlements are likely to be provided with services, and the cost of those services will rise (Gilbert 1990). The overall result will be that more families will occupy smaller plots, will take longer to consolidate their homes, and will be forced to live longer without services. Self-help ownership will offer even less in the way of a desirable home than it does now.

The second possible outcome is that the proportion of families who are renting homes will increase. With real household incomes falling and with formal sector construction being cut back, fewer middle-income families will be able to buy their own homes. Unwilling to live in self-help areas, they will choose to move into rental accommodation. Lower-income households will be deterred from self-help ownership by the rising cost of land, relative to their incomes, and by the increasing difficulty of living in peripheral settlements. Provided that rental accommodation is available, the proportion of tenants will increase, or at least will cease to decline as rapidly as it has in the recent past. The problem with this scenario is that more people are likely to be living in the limited accommodation available. Overcrowding is likely to increase.

The third possible outcome is that access to both home ownership and renting will become more difficult. Home ownership will become too expensive for the reasons already mentioned and landlords will fail to fill the gap. Faced by falling real incomes among the middle class and poor alike, landlords may be disinclined to let accommodation in return for what they consider to be inadequate rents. Alternatively, rental accommodation may become available but at prices that many poor families cannot afford. In either event, neither ownership nor renting is possible. Under these circumstances, the only real option for many will be to share a home; always a fallback position for newly formed households born in the city, but increasingly an inevitability for households where the main breadwinner loses his or her job.

Under current circumstances, therefore, it is unfortunate that more is not known about rental and shared housing. If renting and sharing constitute two of the more likely responses to recession, more ought to be known about how they operate. Admittedly, research on these forms of tenure has become more common in recent years. In Latin America, interesting work has begun to trickle out of Argentina, Brazil, Colombia and Mexico (Portillo 1984, Yujnovsky 1984, Coulomb 1985a,b, Jaramillo 1985, Cuenya 1986, 1988, Necochea 1987, Bonduki 1988, Kowarick & Ant 1988, Gazzoli et al. 1989). In addition, researchers based in London and in The Netherlands have conducted studies in Bolivia, Colombia and Mexico (Edwards 1981, 1982, Gilbert 1983, Hoenderdos et al. 1983, Gilbert & Ward 1985, Beijaard 1986, Green 1988, Gilbert & Varley 1991, van Lindert 1991, van Lindert & van Westen 1991). Similarly, some United Nations agencies and the World Bank have begun to show some interest in the rental housing question (Urban Edge 1984, Mayo 1985, Mayo et al. 1986, UN 1988, UNCHS 1989, Hoffman et al. 1990, Malpezzi & Ball 1991). Rather belatedly, research on this issue has begun to increase, not only in Latin America but also in parts of Africa and Asia (Amis 1984, Rakodi 1987, Tipple 1988, Angel & Amtapunth 1989, Barnes 1987, India NIUA 1989, Amis & Lloyd 1990, Hoffman et al. 1990, Ozo 1990, Willis et al. 1990, Potts and Mutambirwa 1991).

As we study rental housing in more detail, we have begun to discover that some of our conventional assumptions about it do not necessarily hold. Only now are we beginning to recognize how many tenants live in Latin America's cities: well over four million in Mexico City alone. And, while their relative importance has declined, the absolute number of rental units in many places has been expanding (Coulomb 1985a,b). This new accommodation is not where it used to be; the vast majority of new rental housing is located in the consolidated self-help periphery rather than in the centre of the city (Edwards 1982, Gilbert 1983, van Lindert 1991). Similarly, we are learning that the typical landlord is no longer an urban *hacendado* but a much more humble citizen (Edwards 1982, Gilbert 1983, Coulomb 1985a,b). Unlike the stereotypical view of the British or North American slumlord (Daunton 1987), and the reality of large-scale landlordism in a "Third World" city such as Nairobi (Amis 1984, Lee-Smith 1990), few Latin American landlords seem to own many properties.

4

We are learning something, too, about the links between rental and shared housing and changes in the urban land market. There is clearly a positive relationship between the availability of peripheral land and the affordability of rents (Gilbert & Ward 1985, Gilbert & Varley 1991). We are also beginning to discover that tenant households do not inevitably live in slums; indeed, they often live in better housing conditions than owner–occupiers (Lemer 1987). As a result, not every tenant wishes to become a home owner, at least not immediately and not under the current circumstances (Gilbert & Varley 1991). It seems that rental housing can play a positive rôle in the needs of some low-income families, as authors such as Turner (1967, 1968) argued long ago. Finally, our images of different zones within the city are coming into question. As Eckstein (1990) complained recently, "the widely accepted views both of shanty-towns and centre-city slums are no longer applicable. The depiction of peripheral areas is too upbeat and that of the inner city too downbeat." Some rental housing areas offer decent accommodation in desirable locations, some peripheral self-help settlements offer little either in terms of location or in terms of quality of housing.

Despite this recent upsurge of interest, there is still a great deal we do not know about rental and shared housing. This is particularly true of sharing, a greatly neglected area despite some recent attention paid to this issue in Santiago (Ogrodnik 1984, Necochea 1987, Soto 1987). The present book and the research on which it is based represent an attempt to fill some of these gaps in our knowledge. Underpinning that research were a number of specific questions.

First, we wanted to understand how the poor gain access to land in each city, and to establish whether the availability, or otherwise, of cheap land was an important influence on the proportions of families renting and sharing. This was a researchable question in so far as the three cities included in the study contain very different kinds of low-income land market. In Caracas, land invasions have been common for many years; in Mexico City, invasions are seldom permitted, and illegal subdivisions have become the normal method of acquiring land; in Santiago, neither land invasions nor illegal subdivisions have been permitted for nearly two decades, and low-income ownership is available only through subsidized public programmes.

The second aim was to establish who was constructing housing for rent. Were traditional large-scale landlords continuing to let rental accommodation, or had they vacated the market, leaving it to a new breed of self-help landlords? In so far as there were self-help landlords, who were they and why were they constructing housing for rent when so many argue that investing in rental housing is an unprofitable activity?

Thirdly, to what extent were tenant and sharer households part of a poverty-stricken population, constrained from owner-occupation by their inability to obtain land and materials? Do such households want to be owner–occupiers or is this a figment of the government's, and the real-estate lobby's, imagination? Is this a population excluded from the home ownership it craves, or do renting and sharing have certain advantages that sometimes make them preferable alternatives? Is income the critical ingredient in residential choice, or do stage in the life-cycle and the birth of children also influence the process of tenure selection?

Fourthly, given the unfavourable image associated with landlordism and renting, how do landlord–tenant relations operate? Do problems arising from the collection of rents and the failure to maintain the property condemn landlords and tenants to continual enmity? Are tenants constantly on the move due to eviction, or are landlords constrained from evicting even the most undesirable of tenants? When do tenants mobilize to protest about their housing conditions? How many landlords and tenants belong to representative organizations? What effect has government intervention in the housing market had on landlord–tenant relations? How does the law operate, does it operate well or badly, and what use is made of the courts by tenants and landlords?

As I have already suggested, these questions are more than simply interesting avenues for research. They are critical ingredients in improving the quality of government housing policy. Without an adequate understanding of these issues we cannot begin to answer a further series of questions about policy. To what extent do current government policies harm the rental housing sector? Have rent controls damaged the interests of landlords as much as they claim? Should governments encourage investment in rental housing, reversing the trend over the past 30 years towards the growth of owner-occupation? And, if they choose to assist rental housing, what is the most effective way of attempting to do this?

CHAPTER TWO

Methodology

Any comparative study faces a fundamental question: what is the basis for a good comparison? Some comparisons make sense whereas others do not.[1] In general, it seems more sensible to concentrate on similar rather than totally dissimilar phenomena or processes. The principal reason for making a comparison is to shed light on how some common process produces different kinds of results in different places, or to examine why different processes produce similar results. If both processes and results differ, is there much point in the comparison? If, for example, we were to compare the housing situation of New Yorkers with that of Australian aborigines living in the outback, would we actually learn anything very useful about housing? In this respect, I believe that Abu Lughod (1976) is correct when she argues that there has to be the basis for "legitimate comparison" and recommends the approach she used in comparing three North African cities: "looking first at what they have in common, then at the major differences among them, and, finally, at the common processes (applied in variable degrees) which have led to the wide and very real differences which we now find."

A particular problem facing comparative research is the danger of superficiality, and it is difficult to steer a sane path between the desire for generalization and the need for detail. Sufficient detail can be assured only by conducting case studies "in-depth", something which "is difficult to do . . . on more than a few cases" (Ragin 1989). Apart from anything else, too many case studies simply produce too much information to assimilate. Even if the material can be understood by the researcher, it is difficult to convey the necessary details to a reader without writing at far too great a length. Only by exceeding the tolerance of most publishers is it possible to provide enough explanation for the reader to understand

---

1 For a fuller discussion of the problems and advantages of comparative research see Gilbert (1991).

processes at the individual city level while concentrating on cross-city generalizations and comparisons.

Having contemplated these issues, the research for this book was based on a comparative study of three ·cities. The three case studies were selected because they offered the basis for "legitimate comparison"; they were similar, but not all that similar.

Comparing Caracas, Mexico City and Santiago

Caracas, Mexico City and Santiago are all Spanish American cities and therefore they share a number of similarities. First, the nations of which they form part have all been independent for a similar length of time, are governed by not dissimilar administrative systems, and share certain commonalities of culture and language. Secondly, all three cities are relatively affluent; in world terms their countries are firmly placed in the ranks of the middle-income nations and, within Latin America, they can certainly not be regarded as poor (Table 2.1). In 1989, Venezuela was the region's second richest country, Mexico the fourth and Chile the sixth (World Bank 1991). Thirdly, all three countries managed to accumulate huge external debts during the 1970s, a misfortune that has had a major

Table 2.1 Caracas, Mexico City and Santiago: comparative indicators.

	Caracas	Mexico City	Santiago
Per capita income of country in 1989 (US dollars)	2,450	2,010	1,770
Per capita income growth rate 1981–89	-24.9	-9.2	9.6
External debt/exports (%)	227	289	171
Annual inflation 1989	81.0	19.7	21.4
City population 1990 (millions)	3.0	15.8	4.7
Annual growth of city population 1940–80	5.3	5.6	2.9
Annual growth of city population 1980–90	1.5	3.0	na
% of urban population living in city 1990	16	19	44
Population living in rental or shared housing, c. 1980	37	46	36
Minimum wage at time of survey (US dollars)	73	70	47
Unemployment (1989)	4.3	2.9	7.2

Source: UNECLAC (1990); World Bank (1991); respective national housing and population censuses.
Note: The population of Caracas in 1990 was probably underestimated by the census. The figure for Santiago is based on the 1971–82 growth rate. The figures for urban population living in Santiago are for 1982.

impact on their subsequent development experience. In response, and with varying degrees of enthusiasm, each government has moved towards the liberalization of trade, reducing its government budget deficit, privatizing state enterprise and cutting subsidies to the poor. Adjustment policies have been introduced at different times, but everywhere they have had a devastating impact on the urban poor.

The three cities have several important characteristics in common. First, they all have extremely large populations, with Caracas and Santiago in the 3–5 million band and Mexico City competing for the title of the world's largest urban agglomeration. Each is the political capital of its country and each dominates the national urban and economic system.

Secondly, although each city expanded rapidly between 1940 and 1980, each experienced a much slower population growth rate during the 1980s. Slower demographic growth was partly the result of falling fertility levels, but fewer migrants have arrived in the three cities in recent years. Reduced rates of migration were the consequence of economic decline; the result of lower import tariffs, and therefore greater competition for local industry, reduced government spending, export orientation, and a belated response by investors to growing urban diseconomies.

Thirdly, all three cities suffer from extremely unequal distributions of income, a high level of residential segregation and a large number of poor households. In each city, affluent islands of élite housing stand out in the sea of semi-serviced slums and self-help settlements which accommodate the poorer half of the population. Because of inflation and austerity programmes, the poor have recently experienced a very marked decline in their standard of living.

Fourthly, while the authorities in each city have constructed large numbers of public-sector homes, none has managed to reduce the so-called housing deficit.[1] While each of the cities is indelibly marked by government building programmes, there are far larger areas of irregular settlement.

Fifthly, each city has a large rental housing stock, accommodating between one-fifth and two-fifths of the population. In all three cities, this form of tenure has been in relative decline over recent years, a consequence of the preference for owner-occupation which has been the explicit objective of all three national governments. Nevertheless, at least 1 million people occupy rental or shared housing in each of the cities.

Finally, each city has a well developed but overstretched transport system. Each has a modern underground railway whose efficiency stands in marked contrast to the anarchy overhead. Congestion, overcrowding, long journeys to work and air pollution are major problems in all three

---

1 In both Caracas and Mexico City, the public housing stock made up one-sixth of all homes in the late 1970s.

cities. Housing choices are strongly influenced by the transport situation.

Despite the many similarities, there are also important differences between the three cities. At the national level, the political systems are, of course, very different. At the time of the survey, Chile was nearing the end of 16 years of military rule, Mexico had once again elected its president from the all-dominant PRI, and only Venezuela could be claimed to be a democracy (in the sense of a regular exchange of power through the ballot box). These differences in political regime mean that different combinations of populism and repression have been used locally to control the population. The difference has been manifest clearly in the different reactions to irregular settlement.

The three countries also show important economic differences. Both Mexico and Venezuela are major exporters of oil, whereas Chile has to import all of its needs. Inflation has been a problem in all three countries, admittedly at different times: Chile in the early 1970s, Mexico in the middle 1980s and Venezuela in the late 1980s. Similarly, while all three economies have experienced periods of some difficulty, these problems have arisen at different times. Economic recession hit Chile hard in 1972–3 and 1975, and again in the early 1980s; in Mexico recession was very marked between 1982 and 1989; in Venezuela severe problems have been apparent ever since 1979.

The three cities also show some intriguing physical and demographic differences. Certainly, the demography of each city varies. Whereas both Caracas and Mexico City grew very rapidly between 1940 and 1980, Santiago grew more slowly, mainly because national fertility rates in Chile have been much lower for a long time. As a result, there is a clear difference in the structure of the population. In Caracas and Mexico City a majority of the adult population are migrants; in Santiago a higher proportion have been born in the city. In addition, Caracas contains a large foreign-born population; by contrast, there are few immigrants in either Mexico City or Santiago.

Secondly, apart from a common shortage of water, the controls on physical expansion have been different. In Caracas, the main constraint on urban growth has been its topography. Confined to a series of narrow valleys, the city is desperately short of space. In Mexico and Santiago, topography poses less of a problem, and growth has been able to spread at lower densities. The problem in these cities has been how best to service the consequent urban sprawl.

Thirdly, the poor in each city have developed different ways of occupying land. In Caracas, invasions are the principal form of settlement creation, although most families have purchased plots once a firm claim to the land has been established. In Santiago, invasions were the principal means of obtaining land in the early 1970s but the Pinochet government subsequently prohibited them, together with other forms of illegal subdivision. In Mexico City, a variety of methods of land alienation have existed, with the illegal subdivision of private, communal and *ejido* land dominating. These different methods of land occupation have had a significant effect on the evolution of the housing markets in each city.

Fourthly, recent governments have reacted very differently to existing areas of irregular settlement. In Santiago, the authorities carried out a massive programme of eradication and relocation during the 1970s and 1980s. Such programmes have been the exception in Caracas and Mexico City, where gradual servicing has been the most consistent response. In all three cities, however, there has been a common and growing tendency to provide irregular settlement with title deeds.

Finally, there is a different attitude in each city towards housing subsidies. Although in the past all three countries built public housing on a large scale, the 1980s saw major changes in terms of the policy towards subsidies. At the time of the survey, generous housing subsidies were available to many poor families in Santiago, whereas in Mexico City they were available only to victims of the 1985 earthquake, and in Caracas housing subsidies were being phased out.

Methodology

Separate teams worked in the three cities, each following a similar approach and employing a similar questionnaire. Each team included researchers with long experience in the housing field: they knew and understood their own cities and were acquainted with many of the professionals working in government agencies. The method required a delicate balance to be struck between allowing members of each team to follow their own natural interests, and preventing them from diverging too far from the common elements in the research. As co-ordinator of the comparative study, I interpreted my rôle as one of ensuring that similar kinds of information were collected in each city. Unless data collection was carefully controlled, different local situations and different team interests would produce substantially different data sets, making comparison impossible.

We sought to maintain this common thread by using very similar questionnaires and survey methods. Each questionnaire included the same basic questions, and each team was required to work in similar kinds of housing area. Members of each team were also required to answer similar kinds of questions about the housing situation in their cities. At the same time, it was necessary to recognize that each city had certain distinctive features. Our basic approach to this diversity was to allow each team flexibility to introduce individual ingredients into the research design. Thus each team was permitted to focus on particular elements which were of particular importance in their own city. In Mexico City, for example, the production of rental housing was of particular importance to the team,

11

and in Santiago the issue of shared accommodation had become a major political question. The task of co-ordination, therefore, involved allowing each team room to study particular issues while, at the same time, guaranteeing that each team was still conducting the same study as the rest.

The main tools of co-ordination were the three meetings held during the course of the research and the detailed guidance notes which were sent out during the study. The task of preparing such notes was considerably assisted by the fact that I had already undertaken a similar comparative study of rental housing in two Mexican cities (Gilbert & Varley 1991).

The first task for each team was to produce a background paper on the land market in each city. These papers formed the basis for discussion at the first co-ordination meeting. A key objective of the background papers was to identify the main socio-economic and physical characteristics of the cities, and to provide an overview of housing policy in each country, both currently and in the past. The papers were intended more as literature reviews than as original contributions to knowledge. Each paper covered the growth of the city since 1900, the present urban morphology, housing policy since 1950, administrative organization of the city, the main mechanisms through which the poor obtain land, tenure structure, general housing conditions and a resumé of the principal political and economic highlights in the city and the country in recent years.

The documents were used in the process of selecting the settlements where the questionnaire survey would be conducted. Selection was a particularly sensitive issue because it was clear that we could not afford to conduct a complete survey of the whole of each city. As such, it was necessary to choose settlements that were representative of particular kinds of housing situation. The original research proposal suggested that each team should choose three different kinds of settlement: a settlement formed within the past 6 years, which was dominated by recent owners; a consolidated settlement, 20–30 years old, with a roughly equal mixture of owners and renters; and an older, central, rental area, preferably one without too many workshops or too much "gentrification", containing large numbers of poor tenant households.

The choice of these three kinds of settlement was based on the premise that we would be able to compare households in different tenure situations and at different points in their life-cycles and housing careers. We could compare tenants in the central areas with those in the consolidated periphery; the two kinds of tenants with former tenants now establishing themselves as self-help owners in the periphery; and compare the owners of now consolidated homes in the older periphery with their younger counterparts in the new periphery.

However, the process of selection is always less easy in practice than in theory. Each team was asked to prepare a list of two or three alternative settlements which satisfied the criteria for each of these three groups.

Slides of each settlement were brought to the first co-ordination meeting, together with information on the date and history of settlement formation and servicing, the numbers of families living in the settlement, the approximate tenure structure, the level of housing consolidation and servicing, and the location of the settlement relative to the central business district and other major employment centres.

On this basis we proceeded to choose the settlements; a complex process because urban processes in Caracas, Mexico City and Santiago are, in important respects, rather different. Methods of land acquisition, for example, differ considerably. Whereas land for self-help construction has been available in Caracas and Mexico City over the past 20 years, admittedly in more and more distant locations, land invasions and illegal subdivisions virtually ceased in Santiago when the government of Augusto Pinochet took over in 1973. As a result, there are no new self-help settlements on the edge of Santiago. New owner households in Santiago are not self-help builders, they live in mass-produced apartments and benefit from 75% subsidies from the government. In this respect, a direct comparison between Santiago and the other two cities is simply not possible.

Another complication was that Caracas and Mexico City fall into two administrative areas and are governed by different political authorities. This has influenced the form of land alienation and servicing in important ways. In order to understand how these different forms of land alienation affected tenure choice, it was necessary to include settlements from each administrative area. As such, we were forced to conduct interviews in more settlements than we originally planned. In the end, we studied five settlements in each city: one central area, two consolidated self-help settlements and two new peripheral settlements. In the case of Santiago, of course, "new" settlements are much older than the equivalent settlements in Caracas or Mexico City because of the prohibition on illegal land development since 1973.

The next stage of co-ordination was to agree the content of the questionnaire. Each team was sent a copy of the questionnaire employed in Guadalajara and Puebla in the earlier rental study (Gilbert & Varley 1991). The teams tried out this questionnaire on a few households before the first co-ordination meeting. We then made several changes, modifying the questionnaire to accommodate local differences in language and in land and housing practice, and extending the scope of the questionnaire to cover tenants and sharers in more depth and to include a specific section on landlords. The final questionnaire was a little different in each city, but included a large core of identical questions. Each team added some questions but none eliminated any questions being asked elsewhere. In this way we maintained the comparative research element.

After the basic questionnaire had been modified, a common coding guide was produced, intended to simplify later comparison and to prevent

13

differences in interpretation of the data. With a few necessary modifications, each team used this common guide.

Around 700 interviews were conducted in each city. The interviews were conducted with landlords, owners, tenants and sharers; the number of each type of household being listed in Table 2.2.

Table 2.2 Details of the household surveys.

Settlement/City	Type of settlement	Owners	Landlords	Tenants	Sharers	Total
CARACAS		382	41	276	0	699
El Carmen La Vega	Consolidated	104	21	96	0	221
El Carmen Petare	Consolidated	78	20	79	0	177
El Chorrito	New	100	0	1	0	101
Las Torres	New	100	0	0	0	100
San José/San Juan	Central rental	0	0	100	0	100
MEXICO CITY		320	40	240	80	680
El Sol	Consolidated	80	20	80	40	220
Santo Domingo de los Reyes	Consolidated	80	20	80	40	220
Avándaro	New	80	0	0	0	80
Belvedere	New	80	0	0	0	80
Doctores	Central rental	0	0	80	0	80
SANTIAGO		318	47	239	160	764
La Bandera	Consolidated	80	24	79	80	263
San Gregorio	Consolidated	80	21	80	80	261
Eleutorio Ramírez	New	79	0	0	0	79
San Rafael	Old periphery	79	0	0	0	79
Conventillos	Central rental	0	2	80	0	82

Note: In Santiago, the sharer total includes 80 interviews with *allegados*, a special form of sharing household (see Ch. 5).

Reflections on the methodology

Comparative research is not easy and can be conducted in a variety of ways (Abu Lughod 1976, Ragin 1989, Gilbert 1991). A key decision involves whether to use one team or several. In this study it was decided to use local, separate teams in each city. The main advantage of using local teams is that the researchers already know a great deal about their own cities and often have ready-made contacts in the bureaucracy and in the settlements. Being known in the city is a great advantage, at least in those cities where there are not great schisms based on politics, race, religion or ideology. For an outsider, particularly a foreigner, getting to know a city takes time. Clearly it is essential, as in this case, to use researchers with long experience of housing in their own cities. It is especially helpful when they are doing this on a full-time basis and are themselves conducting the field research and doing the bulk of the writing.

However, it would be wrong to pretend that there are no difficulties with this approach. First, the use of separate teams, while clearly bringing major advantages, does complicate the task of co-ordination and can bring conflict. In this project relations both between the teams and between the teams and the co-ordinator were basically very good. On a personal basis, everyone got on very well. Over the duration of the project, however, we had the odd sticky patch. One occurred in establishing an agreed format for collaboration; it might not have been established at all without a number of good meals and the fact that all the people concerned were very amiable. Differences appeared over the methodology to be employed, for example over the balance between a quantitative approach and a more anthropological approach. There were also some differences with respect to how the project should be written up. With respect to policy formulation, we had a lively debate in our final meeting about whether a common policy could be devised for cities as distinctive as Caracas, Santiago and Mexico City. Most amusingly, at least in hindsight, we did not always manage to meet in the same place at the same time; in September 1989 I travelled from London to a co-ordination meeting in Mexico City, only to be told on arrival that the meeting had been cancelled while I was on the plane!

Secondly, the teams did not work at the same speed. In this project, the Santiago team started late, having experienced problems over finalizing the budget, and never caught up. This delay on the part of one team obviously delayed the production of the comparative analysis of the housing situation in each city. In addition, the members of the teams had different levels of commitment to the overall project, largely as a result of their own career developments and interest in the topic.

Thirdly, it would be erroneous to give the impression that the three teams were as interested in the comparative element of the research as in the part that related to their own cities. This attitude arose partly from the differences that are apparent in the housing situation in each city; what is important in one city is not necessarily of importance elsewhere. This encouraged the tendency for each team to want to collect different kinds of information to answer locally significant questions. As I have indicated, we sought to overcome this by agreeing a *modus vivendi* devising a methodology which would collect enough comparable material to guarantee that we could truly compare the three cities, while at the same time allowing each team to collect information on the locally specific. Each team used the same questionnaire, adding extra questions where they wished, but promising not to subtract any. They also agreed to use a similar method for selecting survey settlements, and sampling within them, but they reserved the right, and were given the resources, to pick an additional settlement or two. As a result of this approach, I believe we managed to collect broadly comparable survey data across the three cities.

Despite these kinds of problems, I have no doubt that comparative research of this kind is essential. It is intellectually stimulating because it forces the researcher to eschew simple statements of process, by demonstrating either that in another city the same phenomenon is explained by different processes or that different phenomena are brought about by the same process. Comparison creates confusion, but it is creative confusion. It forces any investigator to think harder than he or she would have done otherwise. Comparison is, of course, no guarantee of good research, but in so far as it tends to discourage the facile conclusion, it is a strong stimulus to better research. If the researcher can avoid the temptation to conclude that all is wholly similar or all is completely different, the full virtue of comparative method is revealed. I only hope that the final shape of this book exemplifies the strength of this approach.

CHAPTER THREE

Housing in Mexico City

Introduction

The development of housing in Mexico City cannot be understood without some background on the political economy of Mexico and the evolution of its capital. This introduction provides a brief description of the country's recent economic development and the evolution of its political system, as well as a review of Mexico City's demographic growth and the way the administrative structure of the metropolitan area has influenced the development of the housing market.

Economic growth and the distribution of income
From 1940 until 1980, Mexico was one of Latin America's economic success stories. During the 1940s and 1950s, the phase of import-substituting industrialization had allowed the Mexican economy to grow annually at over 6%, turning the country into Latin America's fifth most affluent economy by 1960. In the next decade, economic growth continued at a reasonable pace and prices rose annually by only 3% per annum. Of course, Mexican development was severely hampered by one of the most inequitable distributions of income in Latin America, but at least living standards were generally rising in the urban areas (Cordera & Tello 1984). Between 1950 and 1975, real manufacturing wages increased some 2.5 times (Bortz 1984). Welfare indicators, such as infant mortality and school attendance, were improving consistently (Wilkie et al. 1988).

In the 1970s, the pace of economic growth slowed, but Mexico's economic prospects seemingly improved dramatically when major new oil reserves were discovered in 1975. Production increased from 46 million gallons in 1975 to 159 million in 1982 (UN 1988). The rate of economic growth rose markedly; gross domestic product increased at over 8% annually between 1978 and 1981.

Unfortunately, the oil windfall was not spent wisely. A policy of investing heavily in the Mexican economy was based upon increasing

17

public-sector borrowing and contracting large foreign loans. The public sector's deficit increased from 5% of gross domestic product in 1978 to 18% in 1982, and the country's foreign debt increased from US$31 billion in 1976 to $82 billion by the end of 1982. As interest rates were lower than inflation during the 1970s, increasing overseas debt appeared to be a sensible strategy to accelerate economic growth, the Mexican government could repay the debt in the future out of its vastly greater petroleum incomes. Unfortunately, neither the oil boom nor the phase of cheap real interest rates was to last. Problems began to confront the Mexican economy in 1979, and the "debt crisis" broke in August 1982 when the Mexican government admitted that it would be forced to suspend interest payments.

Between 1982 and 1988, per capita gross domestic product fell by 11% (Table 3.1). Inflation rose to unprecedented levels, rising by 99% in 1982 and achieving a spectacular high of 159% in 1987. Government policy put strong emphasis on repaying the debt, with the corollary of cutting back on government expenditure and, hence, subsidies. The effect of the recession on income levels was dramatic. In 1983 the real value of average wages fell by 23%, by 1988 they had fallen to 69% of their 1982 value. The real value of the minimum wage was falling even more quickly; between 1981 and 1989 it had fallen by more than half (UNECLAC 1989).

Table 3.1 Mexico: principal economic indicators.

Year	GDP growth rate	Cons-truction	Oil exports (US $b)	Debt repayments /exports	Inflation	Real minimum wage
1960–69	7.0	7.8			2.5	
1970–79	4.5	(5.4)			16.5	(90.9)
1980	8.4	12.3	9.9	30.2	29.8	100.0
1981	7.9	11.8	13.9	37.2	28.7	101.9
1982	−0.5	−5.0	15.8	47.6	98.8	92.7
1983	−5.2	−18.0	15.5	37.5	80.8	76.6
1984	3.6	3.4	16.2	37.2	59.2	72.3
1985	2.6	2.7	14.7	37.2	63.7	71.1
1986	−3.8	-10.3	6.2	38.3	105.7	64.9
1987	1.8	1.5	8.5	29.7	159.2	61.5
1988	1.4	−3.3	6.5	29.9	51.7	54.2
1989	3.1	2.1	7.9	28.6	19.7	50.8
1990	3.9	7.7	9.4	24.3	30.2	45.5

Sources: UNECLAC *Balance preliminar de la economía de América Latina y el Caribe,* various years; UN *economic survey of Latin America,* various years. *Notes on bracketed figures:* Construction for the 1970s is for 1975–79 only and the real minimum wage is for 1970.

The distributional effects of the recession clearly rebounded on the poor (Martínez 1989). By 1987, it was estimated that 41 million Mexicans, 51% of the total, were living below the poverty line. This represented an increase of 28% in the number of the poor since 1981 (*Latin American Weekly Report*, 31 October 1991).[1] However, it was not only the poor who suffered. Indeed, there is some evidence to suggest that the incomes of middle-class households fell by more than those of the poor, the latter compensating by putting more people into the labour force (INC 1989). If both the poor and the middle class suffered badly, the same cannot be said of the rich. Those in a position to speculate either in stocks and shares or in foreign currency did well during the recession. In 1985, it is estimated that whereas the poorest 40% of Mexicans received only 13% of total household income, the richest 20% received 51% (Hernández & Parás 1988).

If the economic recession had different consequences for different social groups, it also had profoundly different regional impacts. In general, the north grew on the basis of its export-oriented production, whereas those regions dependent on the internal market declined. If Mexico City prospered hugely during the boom, it lost heavily during the recession. It suffered badly from industrial decline, from a higher than average fall in wages and from the rising incidence of crime (Cordera & González 1991). It should be remembered that the household survey in Mexico City was carried out at the end of 7 years of severe recession.

The economic recession seriously affected the construction sector. Although public housing finance programmes were maintained at a high level (see below), the recession had a severe impact on the sector in 1982 and 1983, and again in 1986 (Table 3.1).

The political context

Mexico has been ruled since 1929 by the nominee of the country's most important political party, the Institutional Revolutionary Party (PRI) (Cornelius & Craig 1988). Elections have been held regularly, if not always fairly. Over the years the PRI has maintained a degree of political stability that has been envied in many other parts of Latin America. When economic growth was combined with extensive expenditure on social programmes, its electoral position was maintained easily. The government might be "distant, elitist and self-serving" (Cornelius et al. 1989), it might even be corrupt, but most people realized that they were gaining from economic growth and political stability. Of course, the 1980s undermined one of the basic planks of this success. The recession cut living standards and the PRI came under severe electoral threat during the presidential

---

1 The original figures were supplied by the Comisión Nacional de Alimentación and the Programa Nacional de Solidaridad.

elections of 1988. It is not unlikely that the PRI candidate lost the vote to the PRD although the government machine made sure that he won the election.

Before 1988, the PRI was the government's main mechanism for maintaining political support. An elaborate system of patronage reached down to the poorer ranks of society. Land and servicing policy played an important part in this vital game of winning friends and votes (see below). Since 1988, the methods have changed slightly and more social programmes have been managed directly by the presidency. Patronage has been dispensed through new channels notably the Solidarity programme (González & Escobar 1991, Rojas 1991).[1] In Mexico City, extensive efforts have been made to win back the lost votes of 1988. The poor of the Chalco district in the east have been wooed by the installation of electricity and water, other parts of the city by plans to extend the metro. The distribution of land titles has been accelerated. The results of the Senate and Gubernatorial elections of 1991 suggest that the PRI has been highly successful in regaining political control in Mexico City.

The demographic and physical context

Mexico City gained greatly from the process of industrialization. It also benefited from the growth in state power, since most of the federal bureaucracy lived and worked in the city. Given the rapid rise in Mexico's national population after 1940, the demographic impact on the country's capital was obvious. Table 3.2 shows that the average decennial growth rate never fell below 5.5% between 1940 and 1970. Only during the 1970s were there real signs of a slowing in the city's growth, as industry began to deconcentrate to nearby cities such as Puebla, Toluca and Querétaro. This tendency increased markedly during the 1980s when, according to the census, Mexico City's population rose annually by only 1.4%. With 15.8 million inhabitants in 1990, frequent expressions of fear that the city would have 25 million people by the year 2000 seem to have been exaggerated.

The huge expansion in numbers between 1940 and 1980 was accommodated by a massive rise in the physical area of the city. The total area of the city increased almost eight times between 1940 and 1980 (Table 3.3). Physical expansion of this order was possible because the government only partially implemented its planning regulations. Connolly (1982) estimated that by the middle 1970s, 64% of the city had been developed through irregular subdivision. The process of irregular development was encouraged by the presence of large areas of *ejido* land around the city.

---

1 The Programa Nacional de Solaridad (PRONASOL) was introduced in 1989. It distributes about 2% of the federal budget, co-ordinates the expenditure of certain other government agencies, and manages some of the resources generated by the privatization programme.

Table 3.2 Population of Mexico City, 1930–90.

Year	Population	Annual growth rate
1930	1,049	
1940	1,560	4.0
1950	2,872	6.3
1960	4,910	5.5
1970	8,455	5.6
1980	13,735	5.0
	(12,140)	(3.7)
1990	15,783	1.4
		(3.0)

Source: CENVI (1990) for 1930–80. Garza (1991) for 1990 and bracketed 1980 figures.
Note: Doubts have been expressed about the reliability of the 1980 census. It is claimed that the corrections made to the original figures overinflated the population of the metropolitan area of Mexico City. The figure in brackets is Garza's approximation of the real population for 1980.

Table 3.3 Physical development and density of Mexico City, 1900–86.

Year	Area of city (hectares)	Gross density of city (persons per hectare)
1900	2,714	127
1930	8,609	122
1940	11,750	133
1950	24,059	119
1960	47,070	104
1970	68,260	122
1980	91,211	151
1986	120,819	148

Sources: Urbanized area and gross densities: 1900–70, Connolly (1984); 1980, Iracheta (1984); 1986, Delgado (1988); all cited in CENVI (1990).

The *ejido* is a form of property created under the Mexican Reform Law in 1919 and which was expanded dramatically between 1934 and 1940 by President Lázaro Cárdenas. The law gave the community a permanent right to use national property, giving individual members of the community use of the land. The *ejidatarios* around Mexico City were happy to sell off their land because it had limited agricultural potential.

21

Irregular development was also hastened by the presence of large areas of saline land to the north and east of the city, the desiccated lake bed of Lake Texcoco.

More than 10 million people live on land that has been developed irregularly (Connolly 1988); a form of development that has been a feature of all classes of housing, both high-income areas and public housing have often occupied land irregularly (Azuela 1990). Of course, self-help housing has played a major part in this expansion. Whereas only 2% of the city's population lived in self-help housing in 1947 (Azuela 1990), such housing made up 47% of the city's total housing stock in 1970 (Connolly 1977). Today, "a gross estimate for Mexico City is that around 60% of housing production in recent decades can be attributed to the 'popular sector', basically through irregular settlements" (Azuela 1990).

The administrative boundaries

Since 1930, when Mexico City fell wholly into a single administrative area, its growth has spilled across the limits of the Federal District into the neighbouring State of Mexico (Fig. 3.1). Demographically this shift has become increasingly important and, by 1990, the State of Mexico contained 46% of the capital's total population.

The significance of this expansion for housing lies in the different strategies followed by governments in the two administrative areas. This is of particular relevance to the issue of self-help housing and the use made of *ejido* land. The authorities in the Federal District have generally maintained stricter controls over urban growth than their neighbours. At times, this has encouraged urban development to spill over into the State of Mexico. Between 1953 and 1966, for example, the imposition of severe curbs on land invasions and irregular subdivisions in the Federal District by Mayor Uruchurtu strongly encouraged urban growth in Netza-hualcóyotl. In more recent years, the Federal District has been trying to increase residential densities and has been attempting to control the process of illegal land development. A policy of densification has been vigorously applied by the last three national administrations. It has resulted in a rise in population densities and in the removal of several low-income settlements.[1]

Control and densification has been less apparent in the State of Mexico. Until the 1980s, there was very little effort to control urban development in most municipalities. Although Governor del Mazo introduced a strategy to control urban expansion in the state in 1982, a policy which led to some repression of popular housing developments, a huge amount of expansion has still taken place (Duhau 1988a). Although expansion has certainly

---

1 In the case of Lomas del Seminario in the Ajusco area in the south of the city, some claim that as many as 5,000 families were removed (*Proceso* 7 November 1988).

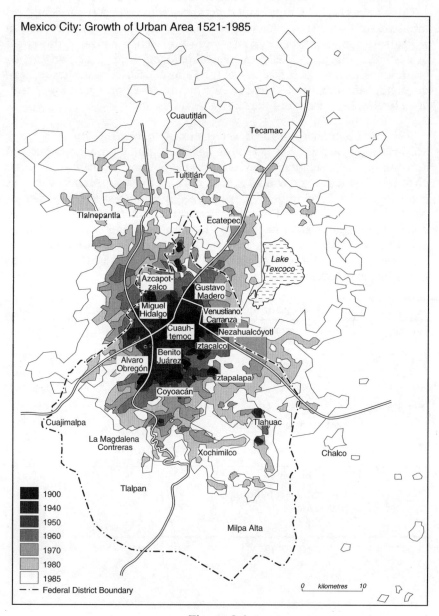

Mexico City: Growth of Urban Area 1521-1985

Cuautitlán

Tecamac

Tultitlán

Tlalnepantla

Ecatepec

Lake
Texcoco

Azcapot-
zalco

Gustavo
Madero

Miguel
Hidalgo

Venustiano
Carranza

Cuauh-
temoc

Nezahualcóyotl

Iztacalco

Benito
Juárez

Alvaro
Obregón

Iztapalapa

Coyoacán

Cuajimalpa

Tlahuac

La Magdalena
Contreras

Xochimilco

Chalco

1900
1940
1950
1960
1970
1980
1985
Federal District Boundary

Tlalpan

Milpa Alta

0   kilometres   10

Figure 3.1

slowed in the north of the metropolitan area, low-density irregular development has continued unchecked in the east (CENVI 1990).[1]

There have also been important differences with respect to servicing and infrastructure provision. Housing in the State of Mexico has always been worse serviced, and poor services were the a basic cause of the social protests in the late 1960s in Netzahualóyotl and Ecatepec. While improvements have occurred since, the Federal District continues to have the better infrastructure. This is shown by the development of the metro, lines having been built only in the Federal District.[2]

Complexity of the housing market in Mexico City

Mexico City's vast size has led to the development of a more complicated housing market than that to be found in Caracas or Santiago. The sheer physical distances over which people have to move in Mexico City has encouraged the development of specialized housing markets in different parts of the city. Different forms of government intervention in the two main administrative areas have introduced further forms of differentiation as have variations in rent control legislation and public housing construction. An attempt has been made in Table 3.4 to describe Mexico City's lower-income housing market.

Table 3.4 Housing submarkets in Mexico City in 1988.

Tenure	Form of housing	Central city	Ring I	Ring II	Ring III
NON-OWNERSHIP	Accommodation with controlled rents	D			
	Free rents	D	S	E & C	E
	Lost cities and rooftop homes	D	E		
	Sharing	E	E	E	E
OWNERSHIP	Public housing		S	S	E
	Earthquake rebuilding programmes	E			
	Self-help		C	E & C	E

Subsectors: E in expansion, C in consolidation, S stable, D in decline.

---

1 The north includes the municipalities of Ecatepec, Naucalpan, Tlalnepantla, Coacalco, Atizapán and Tultitlán; and in the east are the municipalities of Chalco, Iztapaluca, Chimalhuacan and Chicoloapan (CENVI 1990).
2 It was only in 1989 that plans for future lines were announced which would link in the State of Mexico.

Table 3.5 Housing tenure in Mexico City, 1930–90 (thousands of homes).

Year	Owners	Non-owners	Tenants	Total
		FEDERAL DISTRICT		
1930	43	196		239
1940	71	442		513
1950	179	447		626
1960	188	714		902
1970	457	762		1,219
1980	839	908	(729)	1,747
1990	1,166	612	(459)	1,789
		PERCENTAGES		
1930	17.8	82.2		100.0
1940	13.9	86.1		100.0
1950	25.4	74.6		100.0
1960	20.9	79.1		100.0
1970	37.5	62.5		100.0
1980	48.0	52.0	(42.0)	100.0
1990	65.2	34.2	(25.6)	100.0
		MEXICO CITY METROPOLITAN AREA		
1930	43	196		239
1940	71	442		513
1950	183	484		668
1960	226	759		986
1970	656	876		1,532
1980	1,396	1,190	(921)	2,587
		PERCENTAGES		
1930	17.8	82.2		100.0
1940	13.9	86.1		100.0
1950	27.5	72.5		100.0
1960	23.0	77.0		100.0
1970	42.8	57.2		100.0
1980	64.0	46.0	(35.6)	100.0

Sources: CENVI (1990).

Note: Figures are not available on the number of tenants before 1980. The estimates of the total housing stock 1947–80 do not take into account demolitions during the administrative period. Percentages do not always add up to 100% because a number of households did not declare the tenure of their home. This was a particular problem in 1971.

State intervention in rental housing

The origins of rental housing

Throughout the 19th century, the vast majority of families in Mexico City lived in rental housing. This form of tenure continued to dominate the housing stock for much of the 20th century. As Table 3.5 shows, only 14% of the city's population owned their own home in 1940, and even by 1960 only 23% of the population were home owners. It was only during the 1960s that a dramatic change occurred in the pattern of housing tenure. By 1970, the proportion of owner households had increased to 43% and by 1980 it had risen to 64%. Over a period of 20 years the proportion of home owners had increased threefold. And, to judge from recent figures for the Federal District, this trend accelerated during the 1980s.

Before discussing the major factors underlying this dramatic shift, it is necessary to emphasize that until the 1980s rental housing did not decline in absolute importance. Indeed, the number of rental households continued to increase. Between 1930 and 1980, for example, the number of households in non-ownership rose almost ten times. Even when home ownership was expanding so quickly in the 1970s, the number of non-owned units increased by 36%.

Table 3.6 Mexico City: development of rental housing by zones, 1930–80.

Area		Number of houses (thousands)					
		1930	1940	1950.	1960	1970	1980
Central city		178	395	388	472	424	427
	%	90	91	83	86	75	72
First ring		(17)	36	75	233	376	569
	%	nd	69	53	70	50	42
Second ring		–	(9)	(15)	43	65	167
	%	–	45	39	53	36	31
Third ring		–	–	(1)	(6)	(11)	26
	%	–	–	30	43	30	23
State of Mexico		–	–	(17)	(45)	(114)	(282)
	%	–	–	42	54	37	34
Metropolitan		196	442	484	759	876	1,190
area	%	82	86	73	77	57	46

Source: CENVI (1990), based on respective national census volumes.
Note: Figures in brackets were located outside the Federal District at the time of the census. The number of non-owned houses in 1990 in the central ring amounted to 237,885, 48.9% of the total houses in that area. Figures for the outer rings are not available.

Contrary to the impression given so frequently in the Mexican press, Mexican landlords have never ceased to invest in rental housing. What has changed over the years is the location of that rental housing investment and, of course, the social characteristics of the landlords. In the 1930s, rental housing was almost wholly concentrated in what is today the central area. The central *vecindades* increased greatly in number during the 1930s but then production stagnated. After 1940, rental housing was developing more rapidly in the inner ring; in the 1940s, the rental housing stock in this area doubled, in the 1950s it more than trebled.[1] Between 1940 and 1950, the number of new housing units in the inner ring exceeded the absolute expansion in the rental housing stock in the central area for the first time. By the 1960s, rental housing in the central city was in absolute decline and only just maintained its size during the 1970s. By 1980, most rental housing was located in this inner ring and new rental housing was being developed still further out. By 1980, one-quarter of rented and shared housing was located in the State of Mexico.

Over 40 years, therefore, a substantial relocation of the rental housing stock had occurred. In 1940, 89% of the units were located in the central city; by 1980, the share of the central area had declined to only 36%. The central area remained the most typically rental housing area, with almost three-quarters of the households in the central area living in non-ownership, but its relative importance compared to the inner ring fell. By 1980, only half of Mexico City's rental housing was located in this area. No longer was living in rental accommodation synonymous with living in the centre of the city. Rental housing had developed strongly throughout the suburbs, in both the formally constructed and the self-help settlements. In the 1980s, the effects of the earthquake and subsequent government policies have dramatically reduced the amount of rental housing in the central area.

Urban renewal

Urban renewal was an important cause of the absolute decline in rental housing in the central city after 1960. Of course, urban improvement is hardly a new phenomenon in Mexico City for Chapultepec Castle and the great Paseo de la Reforma were built during the reign of Maxmilian (1864–67). But modern attempts at renewal have been more destructive of the housing stock because they have been implemented in more densely populated areas.

The modern schemes began in the 1960s with several spectacular attempts to redevelop the road pattern in the central area. Major road schemes included extending the Paseo de la Reforma to the north,

---

1 Unfortunately, the Mexican Housing Census did not distinguish between renting and non-ownership until 1980.

through a densely populated rental housing area which had developed at the turn of the century. The road improvement programme helped reduce the rental housing stock by some 48,000 units (CENVI 1990). The gradual expansion of the central business district into *vecindad* areas destroyed more housing during the 1960s, as did construction of the Nonoalco Tlatelolco public housing complex to the north of the city centre.

During the Echeverría *sexenio* (1970–76) a major effort was made to move poor families living in the central city into new public housing in the periphery. A campaign was launched against the so-called "lost cities" and against decaying *vecindades*.[1] Families were moved in large numbers to new public housing complexes beyond the central city (CENVI 1990). Thousands of rental housing units were also destroyed by the subsequent administration through a further programme of road building. Major routes into the central area were widened into four- or six-lane highways (the so-called *ejes viales*). These initiatives stimulated a considerable amount of protest from tenant groups but despite this extensive opposition some 50,000 rental housing units in the central area disappeared during the 1970s.

The recession of the 1980s would, no doubt, have slowed the process of urban renewal. Commercial land prices plummeted in real terms between 1980 and 1984, making redevelopment schemes much less attractive (CENVI 1990). However, the process of urban renewal was reinvigorated by a wholly different and unwelcome source; the earthquakes that hit the central area in September 1985. Because of the nature of the subsoil, the central area of Mexico City was very badly affected by the tremors. Ironically, the earthquake almost succeeded in achieving something that many landlords had tried to do for years; remove the tenants so that they could redevelop their property. However, while the earthquakes severely damaged much of the housing stock, the owners were prevented from rebuilding for higher-income groups by the considerable tenant protests that developed after the tremors. Rather than rehousing the affected households in the periphery, the traditional form of response, the government decided to acquire the damaged property through compulsory purchase and to rebuild 45,000 housing units in the central area. Helped by a major subsidy from the World Bank, the government launched an enormously effective programme to produce low-cost homes in the central area. The existing population would stay in the centre, although, ironically, not as tenants. For reasons discussed below, the new homes were sold to the former tenants at highly subsidized prices.

---

1 The "lost" cities were areas of very poor-quality housing created in the inner city on vacant plots of land. The owners of the land either rented out plots or tolerated the occupiers temporarily until they could be displaced (Sudra 1976).

In 1987, the programme was partially extended in a second phase, and, in 1988, a new strategy developed to cater for those tenants living in old property not directly damaged by the earthquake.[1] Neither of these follow-up programmes was nearly so successful, mainly because the funding afforded them was much less generous. With the memory of the earthquake fading and tenant organization becoming less vehement, there was much less pressure on the government to take "radical" action (Connolly 1987). Overall, however, the earthquakes, and the government's subsequent response, had managed to prevent the displacement of low-income households. At the same time, the shift from renting to ownership had accelerated; as shown dramatically in Table 3.5.

Rent controls and rental legislation

In 1942, controls were introduced on property with rents of less than 300 pesos; this measure was extended indefinitely in 1948. There can be little doubt that controls have had a negative effect on the amount of rental housing being produced in Mexico City, but most accounts have certainly exaggerated the impact (Aaron 1966, Grimes 1976). Connolly (1982) argues that the main effect was to slow the pace of land-use change because the controls complicated the transfer of low-cost housing into more profitable uses.

Whatever the precise effect, the results of the legislation were confined to the central area of the city. New property entering the rental housing market beyond the central area was soon attracting rents above 300 pesos and was therefore exempt from the legislation. By 1961, the rent controls affected only 22% of rental accommodation in the Federal District and, by 1976, only 1% of all homes in the metropolitan area were covered by the legislation (SHCP 1964, Connolly 1977).

Despite many proposals in the legislature, no further rent controls were introduced in the Federal District between 1948 and 1985 (Coulomb 1985b). Only the rapid inflation of the 1980s finally elicited some modifications to the Federal District's Civil Code. In 1985 rent increases were limited to the equivalent of 85% of any rise in the minimum salary. Security of tenure was also increased for the tenant and the Federal Attorney for the Consumer was instructed to give tenants free advice, to listen to their complaints, and to act as conciliator between landlords and tenants. A new kind of magistrate (the *Juzgado del Arrendamiento Inmobiliario*) was established who would be concerned only with rental property.

The new legislation angered landlords, especially when the rent controls were later reinterpreted in a way that clearly favoured the tenants. However, since the introduction of the Solidarity Pact in December 1987,

---

1 The *Casa propia* programme.

the rent controls seem to have been ineffective. Since rents, unlike most other kinds of prices, have been excluded from the controls, they have risen more rapidly than the retail price index. After 20 years when rents rose more slowly than other prices, the period since 1988 has seen a considerable rise in the real value of rents (Gilbert & Varley 1991). During 1988, rents increased two-and-a-half times faster than prices and almost three times faster during 1989 (CENVI 1990).[1,2]

If rent controls have attracted most attention, the authorities have also taken other forms of action. *Vecindad* improvement programmes were introduced by INDECO and INFONAVIT in the 1970s and slum improvement was supposed to be a major goal of the Federal District's Directorate for Popular Housing. Some efforts have also been made to encourage owners to build accommodation for rent. During the 1950s, an exemption from local land taxes for 20 years was introduced for owners constructing "popular housing" for rent. However, since land taxes were rather low at the time, this was not a significant incentive.

Greater efforts have since been made to convince the private sector to construct more rental housing. In 1980, the Federal government offered seemingly generous tax-relief certificates to companies producing "social-interest" housing for rent. As this incentive did not stimulate much investment, more attractive conditions were announced. FOVI began to offer builders accelerated depreciation allowances on tax, as well as loans for up to 70% of the cost of housing construction at a highly subsidized rate of interest, providing that the accommodation was let for a minimum of 10 years and that the rent did not exceed the minimum salary. In 1985, the minimum period was reduced to 5 years and tax relief equivalent to 15% of the total value of the investment offered. Despite these incentives there is little sign that companies have rushed to build rental housing in Mexico City (Gilbert & Varley 1991).[3]

During the 1980s, too, the Federal District government (DDF) was required to increase the supply of rental housing in the city. Unfortunately, no additional resources were made available so little happened (Coulomb 1985b).

In general, therefore, government in Mexico has done little to encourage

---

1 The Federal Attorney agreed to raise rents only in line with the January rise in the minimum salary, and not in line with any other changes that might occur during the year. Since the minimum salary was being modified at least twice a year, this clearly undermined the intention of the legislation.

2 During 1990 the rise in rents was 40%, compared to 30% for prices generally, and by early 1991 rents were again rising less rapidly than other prices. In comparison to the minimum salary, rents rose dramatically during the period from 1 December 1987 to 1 January 1991. Whereas the former increased 2.1 times during the period, the latter increased 3.8 times.

3 There is evidence of building being stimulated in major tourist resorts such as Cancún and Puerto Vallarta.

rental housing and has done much to discourage it. In the light of its extensive efforts to encourage the development of private home ownership, it is not surprising that the relative importance of rental housing in Mexico City plummeted after 1950.

Official encouragement for owner-occupation

The private sector

Mortgages to buy new property have been available to middle-class households for many years; and, as in other parts of Latin America, huge new areas of real estate have been developed to satisfy the demand for new suburban housing. The typical alliance of financiers, builders and real-estate developers have convinced the government that suburban development is both desirable and necessary. The government has responded by providing infrastructure and services and by providing tax relief on mortgages. With house and land prices rising, few middle-class families could afford not to buy their own home.

By contrast, the purchase of a home by the poor has been more difficult. Formal sector finance for the private low-income housing market only became available in 1964 with the establishment of the Housing Finance Programme (PFV). With funds from the United States, under the auspices of the Alliance for Progress, and with compulsory lending required of the banks, a major effort was launched to increase the supply of "social-interest" housing (Ward 1990b). The establishment of two funds within the Bank of Mexico, FOVI and FOGA, was highly successful in generating more housing for the middle and working classes; between 1965 and 1970 some 76,000 units were built and, in the 1970s, 107,000 units. Although much of this housing was built in Mexico City, most of the accommodation was too expensive for poor families.[1] The contemporary slogan, "Everyone an owner", proved to be more than a little misleading.

The problem of unaffordability increased after 1972, when prices of the cheapest form of social-interest housing began to rise rapidly relative to salaries; more than doubling between 1973 and 1979 (Schteingart 1990). Investment in social-interest housing fell markedly and has never recovered since. The situation became worse during the recession because few in the city could afford to buy the new accommodation. Mortgage repayment rates were rising fast, while the property was quickly becoming less desirable as design and space standards were lowered. Undoubtedly, the 1980s represented a major crisis in the housing system; a system still based on the aim of increasing the level of owner-occupation (CENVI 1990).

---

1 Between 1963 and 1972, 56% of the housing was built in the metropolitan area.

Public housing

The Mexican government first began to build public housing in 1925. Until the 1950s, however, the few homes constructed went predominantly to government workers. During the 1960s, as we have seen, most housing for lower-income groups was built by the private sector. With the increasing cost of this social-interest accommodation, pressure from the Mexican Workers' Confederation (CTM) persuaded the new administration of Luis Echeverría to increase formal-sector production by making use of national pension funds. Three major housing institutions (INFONAVIT, FOVISSSTE and FOVIMI) were established to build housing for private-sector workers, state employees and the military respectively. The huge expansion in funds allowed annual housing production to increase from 19,960 units in the 1965–70 period to 56,783 between 1971 and 1979 (Garza & Schteingart 1978, Ward 1990b). During the decade, the three new housing institutes built 80,000 housing units in Mexico City, and two local-government organizations constructed a further 61,000.[1] Unfortunately, the total amount of "social-interest" housing being constructed in the capital was falling. The private sector could not afford to build homes for the poor at a time when the latter were increasingly unable to pay the rising costs of the accommodation. Similarly, financial pressures on the public-housing agencies were forcing their programmes up-market.

Table 3.7 Houses constructed in Mexico City, 1947–88.

Period	National total	Mexico City	Public housing/total housing construction in Mexico City
1947–64	121,200	76,894	12.8
1965–70	119,759	39,500	12.1
1971–75	568,523	94,133	17.8
1976–80		88,786	16.8
1981–88	1,084,230	239,648	17.2

Sources: Figures for Mexico to 1970 are from Garza & Schteingart (1978); from 1971 they are calculated from Ward (1990b) and Gilbert & Varley (1991). Figures for Mexico City to 1975 are from Connolly (1977); 1976–88 from CENVI (1990).

With the onset of the recession, the government of President de la Madrid decided to use construction as a means of keeping the level of unemployment as low as possible. The newly nationalized banks were

---

1 DDF built 40,000 units in the Federal District and INDECO constructed a further 21,000 houses (CENVI 1990).

required to invest 3% of their reserves in housing construction. Despite the recession, more homes were built than ever before; it is claimed that more than 1 million houses and apartments were produced nationally between 1981 and 1988 (Table 3.7).

Initially, Mexico City received relatively little benefit from this expansion because official policy aimed to build only 15% of all housing units in the country's three largest cities. The earthquakes of 1985, however, substantially modified this policy. While 61,000 public housing units were built between 1983 and 1985, in the next 3 years the total rose to 149,000. Of this massive number, the earthquake reconstruction programme accounted directly for 54,000 units.

The new housing in Mexico City differed, however, from that being constructed elsewhere, in so far as most of the units were in the form of finished dwellings. Whereas in the rest of the country "progressive housing" had become a priority, land costs meant that this policy was difficult to implement in Mexico City. FONHAPO, which had the lead rôle in this form of housing solution, undertook relatively little construction in the capital. As a result, while annual public housing production was much higher in the capital in the 1980s than in the 1970s, this did not benefit most of the poor. With housing subsidies generally being cut, the major public-housing agencies could only maintain their financial health by cutting the amount of "social-interest" housing in their production schedules. Since building costs were also rising relative to average salaries, access to public housing for the poor certainly deteriorated; increasingly it was the impoverished middle class who were moving into "social-interest" housing.

The beneficiaries of the Earthquake Rebuilding Programme were exceptions to this trend. Because of political pressure, the government was persuaded to offer large subsidies to the population in need of rehousing. While some middle-class families also benefited, large numbers of poor households received subsidies (Azuela 1987).

If there have been considerable fluctuations in public housing policy in Mexico City, the authorities have been remarkably consistent in one respect. While they have built large numbers of homes over the years, very few units have been intended for rent. Indeed, it was only in 1949 that two state institutes began to build accommodation for rent. By 1963 some 18,000 units had been constructed, mainly in Mexico City. In hindsight, the programme was a major failure because neither agency found it easy to collect the rents or to raise the level of the rent during periods of inflation. When both began to lose large sums of money, a decision was taken not to let housing in the future.

This policy appears to have been maintained through the years; indeed, the difficulties of renting public housing seem to have been ingrained in the minds of Mexican officials. Agreement is total that public housing should be sold rather than let. The existing rental housing has been gradually sold off and no more has been built (Gilbert & Varley 1991).

The only rental accommodation that now exists in the public housing sector is that which is being rented illicitly by the owners. The state housing agencies prohibit rental contracts while the purchaser is still paying off the mortgage. Despite this prohibition, it is clear that large numbers of "owners", unable to pay their mortgage, are renting out their property to higher-income groups (CENVI 1990). Tenants in public housing have to enter through the back door.

Self-help housing

Despite the high rate of construction of public and "social-interest" housing during the 1970s and 1980s, only a minority of families could be accommodated. The slack was largely taken up by the self-help housing sector, which expanded enormously after 1940. Most families now live in settlements founded through irregular development and built predominantly through self-help methods of construction. Nevertheless, the supply has rarely been able to keep up with demand. Only during the 1970s has the housing stock increased more rapidly than the population.

Self-help housing has developed in Mexico City on land that was incorporated irregularly, and often illegally, into the urban area. Indeed, half of all the land that became part of the urban area after 1940 suffered from some juridical problem (CENVI 1990). In the past, most housing had been developed on private land, but since 1940 there has been a dramatic increase in the amount of communal or *ejidal* land being used. After 1940, many *ejidos* on the edge of Mexico City have been invaded or have been subdivided by the *ejidal* community itself. The use of *ejidal* land for urban housing is clearly illegal; it has occurred only because of connivance between the *ejidatarios* and officials in the Ministry of Agrarian Reform. Despite this situation, it is estimated that between 1940 and 1975 some 21% of the city's expansion occurred on *ejido* land and a further 27% on common land (Schteingart 1983); in recent years the use of *ejidal* land has become even more marked (Varley 1985).

Other forms of irregular land development contributed to urban expansion. A major contribution was made by illegal land developers who subdivided land without providing the required infrastructure and services. Large areas were developed in this way, most notoriously on land to the east and northeast of the city. During the 1950s, 34 companies sold 160,000 plots covering some 62km^2 of land illegally in the Netzahualcóyotl area alone (Guerrero et al. 1974, Gilbert & Ward 1985). "Not until the 1970s was any serious attempt made to penalise the developers – a consequence of the extensive complicity between state authorities and the companies throughout most of this period" (Gilbert & Ward 1985). Invasions have also been employed to establish new self-help housing areas, although they have been much less important as a form of land occupation than in Caracas or in Santiago prior to 1973. As we have observed, they have been particularly rare in the Federal District, where the authorities have normally taken a hard line against them (Cornelius 1975, Gilbert & Ward 1985). Finally, a few settlements have been established on rented land. With little security, poor households living in these areas were reluctant to invest in housing improvement. Such areas failed to consolidate and became known as "lost cities" (Sudra 1976).

Official intervention in the self-help areas has been intermittent. Clearly, the political authorities have been constantly involved in the process of development in failing to apply either the spirit or the letter of the law. Government officials and politicians have led land invasions, turned a blind eye to illegal subdivisions, and manipulated the regulations to legalize settlements formed on *ejidal* land. The long and fascinating history of the politics of land has been explored extensively in the literature (Cornelius 1975, Eckstein 1977, Perló-Cohen 1981, Connolly 1982, Iracheta 1984, Gilbert & Ward 1985, Varley 1985, Azuela 1989, Ward 1990a).

Few self-help areas were initially provided with services. Invasion areas were not supplied, on principle; and illegal subdividers failed to install infrastructure, in order to guarantee a good return on their investment. Official intervention has always followed a long way behind the needs of the inhabitants. While technical criteria have not been disregarded, the distribution of services has been manipulated regularly for political reasons (Ward 1981, Gilbert & Ward 1985).

Similarly, the legalisation of land holding has been used as a means of maintaining and winning political support (Varley 1985, Azuela 1989). Such was not always the case, for regularization of land title was rare in the Federal District between 1953 and 1966 and only became commonplace after Mayor Uruchurtu left office. It became increasingly important in the late 1960s, when the level of social protest mounted, and several agencies were created in the early 1970s to handle the issue. In the State of Mexico, the political crisis that culminated in a strike against making payments to the subdividers in Netzahualcóyotl led to an official clamp-down on the subdividers and a major campaign to legalize land holding. Between 1969 and 1981, some 350,000 titles were distributed by the State of Mexico authorities, a further 330,000 in the Federal District, and by 1982 around 80% of settlement on *ejido* and community land had been regularized by CORETT (Stolarsky 1982, Iracheta 1984, CENVI 1990). The distribution of land titles became a high priority and was stepped up under the Salinas administration, which announced that 90,000 titles would be issued in the metropolitan area in 1990.[1]

An attempt to control the future process of self-help development arose

---

1 Reported in *Metrópoli*, 6 January 1990.

hand in hand with regularization. Approval of the General Law of Human Settlements in 1976 unleashed a "compulsive wave of urban and spatial planning" which brought with it a stronger effort at controlling irregular settlement (CENVI 1990). Indeed, President López Portillo began his period of administration (1976–82) by announcing that not one more land invasion would be permitted in the Federal District. Fences were established around the land most likely to be invaded, together with warning notices. During the 1980s, an absorption zone of 16,500 hectares was designated in the south of the city, which prohibited low-density residential development. In practice, high-income settlements were permitted to expand but, especially in the Ajusco area, self-help settlements were removed. Many families were displaced, including between 1,000 and 5,000 families during the removal of a large illegal settlement in Lomas del Seminario in November 1988.[1] The level of repression was sufficiently effective that local organizations moved towards a new strategy of organization based on obtaining land "inside the law" (Coulomb 1985, CENVI 1990).

In the 1980s, the policy was applied more firmly, even in parts of the State of Mexico. In 1982, Governor del Mazo announced a programme to restrict urban sprawl. Each municipality was required to designate its urban limits, and settlements developing beyond these limits would be removed. This policy was certainly applied in places, for example 28,000 people living in three settlements in the municipality of Ecatepec were removed between 1985 and 1987 (Duhau 1988a). However, it is clear that whereas the policy was applied actively in the north of Mexico City, it did little to prevent the vast sprawl of new low-density settlement occurring on *ejido* land in the Chalco Valley to the southeast.

Overall, therefore, the 1980s have seen growing signs of planning. Efforts have been made to increase residential densities within the metropolitan area and to control the spread of irregular settlement. The policy has been effective in large areas of the city, although it has not prevented low-income sprawl in the east of the city in zones deemed to be "unsuitable for urban development" (Duhau 1988a). Once established, political priorities have dictated that these unsuitable areas should be serviced. In 1989, 2 million people in the Chalco Valley received electricity and water under President Salinas' Solidarity programme; a clear, and ultimately successful, effort to win back their votes after the PRI's electoral debacle of 1988.[2]

Any sign of planning the layout of self-help areas before they have been occupied has been far less evident over the years. Official sites-and-

---

1 Official accounts say 1,000 families, local organizations argue that the total was 5,000 families (*Proceso*, 7 November 1988).
2 The government was very successful in the Senate elections of 1991.

services programmes have been both infrequent and on a small scale. In the State of Mexico, AURIS introduced progressive housing schemes in both 1971 and 1977, and the Federal District administration launched a programme in 1976 which produced 7,155 "housing actions" by 1982. However, most finance has always gone either into the upgrading of existing settlements or into the building of finished homes (CENVI 1990). The major housing institutions have only recently taken up "sites and services" in a big way. In particular, FONHAPO has espoused this policy on a large scale since 1981. However, few of FONHAPO's schemes have been located in the national capital. It managed to finance only 13,000 units in Mexico City before 1987, only stepping up its programme to around 11,000 units in 1988 (Duhau 1988b).

CHAPTER FOUR

Review of the
Mexico City survey results

Settlement selection

The questionnaire survey was conducted in five settlements in Mexico City. Choosing five settlements in a metropolitan area with around 15 million people was highly problematic. Eventually, one central area, Doctores, was chosen, two consolidated settlements, Santo Domingo de los Reyes and El Sol; and two new settlements, Avándaro and Belvedere. All five settlements are located in the southern half of the city (Fig. 4.1). Given the different forms of settlement formation in Mexico City, it was necessary to select one neighbourhood formed by invasion and another formed through illegal subdivision within each of the new and consolidated categories.

Avándaro is administratively part of the municipality of Chalco, in the State of Mexico. It is located on the southeastern edge of the city, 22km along the Mexico City to Puebla motorway. It was founded in the early 1980s as a result of the irregular sale of *ejidal* land. The area is subject to severe environmental problems, suffering from dust-storms in the dry season and floods in the wet. In 1985, 850 families lived in the settlement, although only about half had lived there for more than 1 year, and one-third of the 1,200 plots were still empty. At the time of the survey, the housing conditions of many families were still fairly rudimentary, only three houses in five had been built out of cement blocks and only 1 in 20 had more than one storey.

The settlement had few services. Water was supplied by tanker, electricity was stolen, and most families were forced to use latrines. The roads were unpaved and the only community facilities were a small market, a nursery and a surgery. In terms of tenure, 78% of the households were owners, 11% had been lent the house and 5% were tenants. In general, living conditions were poor.

Belvedere is in the Federal District within the *delegación* of Tlalpan. It is situated on the flanks of the Ajusco volcano, 18km from the *zócalo*, on the

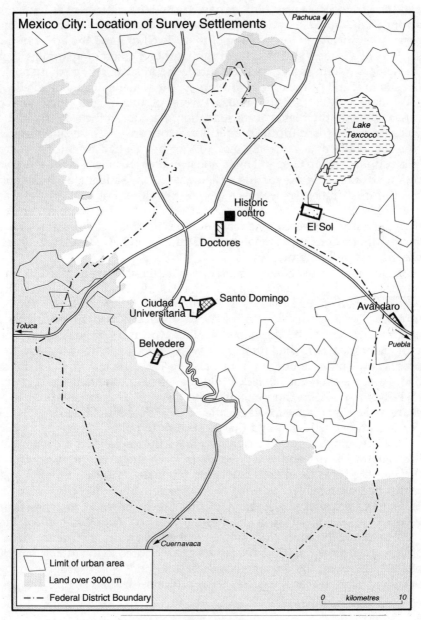

Mexico City: Location of Survey Settlements

Pachuca

Lake
Texcoco

Historic
centro

Doctores

El Sol

Ciudad
Universitaria

Santo Domingo

Avándaro

Toluca

Puebla

Belvedere

Cuernavaca

| Limit of urban area |
| Land over 3000 m |
| Federal District Boundary |

0   kilometres   10

Figure 4.1

southwestern edge of the city. It was formed in 1978 through the illegal subdivision of land which was in dispute between the former *hacendados* and the *ejidatarios*. The authorities have treated the settlement as if it were an invasion. In 1980, various attempts were made to displace the new occupants, including the burning of their dwellings and the arrest of the settlement's leaders. In 1982, the neighbourhood association arranged for the settlers to return and, in 1984, succeeded in suspending payments to the subdividers. At the time of the survey, the settlement contained about 600 houses. Most of the houses were occupied, but they varied considerably in their level of consolidation. Because of the slopes and the volcanic rock, it is difficult to build infrastructure. As a result, services were generally deficient, although 90% of homes had electricity, water was supplied by tankers, and most households used cracks in the volcanic rock for drainage. However, the settlement did have adequate schools and health-care facilities, as well as a CONASUPO store. Public transport was provided in the form of *peseras*, which are quicker but more expensive than the buses that reach other communities in Ajusco. At the time of the survey, there were approximately 2,900 families in Belvedere, virtually all of whom were home owners.

El Sol is a consolidated settlement, administratively part of the municipality of Nezahualcóyotl in the State of Mexico. It is situated on the desiccated bed of Lake Texcoco, some 10km from the *zócalo*. It was formed through illegal subdivision in the 1950s, indeed it was one of the first neighbourhoods to be developed in this way in Nezahualcóyotl. The settlement has long had problems with servicing and, although the process of legalization began in this area in 1973, the majority of families living in the settlement still lack a title deed. In 1988, the household count revealed that 53% of households were owners, 22% tenants and 23% sharers. We suspect, however, that the proportion of tenants is 10 percentage points higher and that of sharers 10 points lower.[1]

Santo Domingo de los Reyes forms part of the *delegación* of Coyoacan in the Federal District and is located 10km from the *zócalo*. It was formed through the invasion of land belonging originally to 1,048 *comuneros* and which had been subject to irregular occupation from the early 1950s. In 1971, this land was the location of the largest invasion that has ever occurred in a single day in the Federal District. With the help of the CNOP, an organization that links popular urban organizations to the official government party, 4,000 families participated in the establishment of the settlement. A few months later the land was expropriated by presidential decree and regularization of the settlement began. Despite

---

1 In practice many landlords did not wish it to be known that they were renting out accommodation in case that would become known to the authorities. Many tenants described themselves as sharers, under instruction from the landlord.

this strong level of official support, some plots still lack legal title. In terms of services, electricity was supplied officially in 1976 and most of the roads are now paved. Communications are excellent, with a metro station located on the western edge of the settlement. However, drainage remains a problem, with one-quarter of the homes using septic pits and the rest relying on natural cracks in the volcanic rock. The houses in the area are now completely consolidated and there is a high population density, 653 persons per hectare. In 1980, 16,500 families lived in the settlement, occupying 9,185 plots. Our census of the settlement revealed that 60% of households were owners, 12% tenants and 28% were sharing. We suspect, however, that many of the sharers were, in practice, tenants.

Doctores was chosen as the site for the survey of central tenants. The settlement is located in the Cuauhtémoc *delegación* in the Federal District, less than 4km from the *zócalo*. The neighbourhood dates from the turn of the century, when it was developed for middle-income housing. After the Second World War many houses were demolished and replaced with apartment blocks intended for rent to lower–middle-income families. Mexico City's general hospital was built on the southwestern fringe of the settlement. From the early 1970s, commercial and office activities began to develop in the settlement and the residential population began to decline; by 1985, less than two-fifths of the area was devoted to housing. This trend was profoundly affected by the earthquake of 1985, when many offices were destroyed and some businesses decided to move away from the area. The tenure structure was also modified by the damage inflicted on the area's *vecindades*. Although 121 *vecindades* were rehabilitated or reconstructed under the Earthquake Renovation Programme, the property was sold to the inhabitants.[1] As a result, 4,222 households (30% of the resident families) changed tenure from tenancy to ownership.

The nature of owners, tenants and sharers

Owners have larger families than non-owners, having on average one additional child. Heads of household are more likely to be migrants than most tenants. They are much older than non-owners, except in the newer settlements where owners are often very young. Owners generally do not earn more than tenants but there are wide variations between owners in the different settlements.

Tenants tend to be younger and have smaller households than owners. There is a higher proportion of incomplete households, particularly of female-headed households, in this tenure group. However, there is much variation between settlements, particularly between the central city and

---

1 See previous chapter on the urban renovation programme.

41

the consolidated periphery. Tenants in the central city tend to be much older and more prosperous, those in the periphery younger and poorer.

Sharers have smaller households and tend to be younger than other heads of household. They have more small children, and few families contain old people. Incomes are lower and they have fewer personal possessions than other groups. Few migrants become sharers, who are twice as likely as owners to have been born in Mexico City. Sharing is a form of tenure for young households beginning their residential career.

The differences between the tenure groups should clearly not be exaggerated, since each group shows considerable internal diversity. The characteristics of owners in one settlement are often more like those of tenants in another settlement than like those of owners elsewhere. Such diversity is also seen between the tenants, particularly between those living in the central areas and those in the periphery.

Income and tenure

Table 4.1 shows that although owners have higher household incomes than tenants, their per capita incomes are lower. In terms of ownership of consumer durables, there is little difference between them (Table 4.2). Sharers have much lower household incomes than either owners or tenants, but they are better off per capita than the average owner. They have a little less than the other groups in the way of consumer durables.

These findings suggest that income in Mexico City is not the key factor

Table 4.1 Household income by tenure and settlement, Mexico City (thousand pesos).

Settlement	Income		
	Household	Per capita	Male
Owners	545	91	379
Avándaro	477	79	367
Belvedere	482	78	325
El Sol	644	106	418
Santo Domingo	581	104	406
Tenants	502	115	393
El Sol	461	96	375
Santo Domingo	513	134	430
Doctores	533	118	374
Sharers	435	101	372
El Sol	467	112	402
Santo Domingo	402	90	337
Total	524	101	395

Source: Mexico survey.

determining differences in tenure among the poor. This point is underlined by the differences apparent within tenure groups. For example, among owners there are major differences in income level by settlement. Owners in the consolidated settlement of El Sol have family incomes almost 50% higher than those of new owners, and per capita incomes 40% higher. Similarly, there are major differences between the tenants: household incomes vary much less than those of owners, but per capita incomes are very different; tenants in El Sol are poorer than those living in the central area and much poorer than those in Santo Domingo.

Table 4.2 Wealth indicators by tenure group, Mexico City (percentage with article).

Article	Owners	Tenants	Sharers	Total
Radio	93	92	94	93
Colour television	12	17	14	14
Sewing machine	27	21	16	24
Refrigerator	49	50	45	49
Car	11	7	5	9
Bicycle	9	7	4	8

Source: Mexico survey.

Employment and tenure

Table 4.3 shows that there are few significant differences in terms of employment between the different tenure groups. Tenants have the highest proportion of males in paid work, but the difference is not large. The proportion of females in paid work is similar between tenure groups. In terms of the nature of their employment, many owners work in the construction industry but very few sharers. Again, variation within tenure

Table 4.3 Employment by tenure, Mexico City

Variable	Owners	Tenants	Sharers	Total
Male employment in industry	21.3	23.0	19.7	21.7
Male employment in construction	17.0	13.1	4.5	13.3
Males in paid work (%)	94.4	97.9	92.8	95.0
Females in paid work (%)	27.9	24.7	29.1	27.0
Self-employed males (%)	24.1	21.4	30.3	26.1

Source: Mexico survey.

43

groups is important; few central tenants are engaged in construction, whereas tenants living in the consolidated settlements are just as likely as owners to be engaged in the building industry. In general, therefore, there are few differences in the employment characteristics of the different tenure groups.

Age and family structure

Owners are much older than tenants, and tenants much older than sharers. However, Table 4.4 shows that there is considerable variation within the tenure groups. Owners in the new settlements are much younger than owners in the consolidated settlements, and tenants in the central city much older than other tenants.

Owners generally have much larger households than tenants or sharers and have, on average, one more child. Extended families are common among all tenure groups but, perhaps surprisingly, are most common among the tenants. This phenomenon is a function of the high proportion

Table 4.4 Age and family structure by tenure, Mexico City.

Variable	Owners	Tenants	Sharers	Total
Number of persons	6.0	4.4	4.3	5.2
Number of children	3.4	2.5	2.2	2.8
Size of nuclear family	5.7	3.9	4.0	4.8
Age household male	42	35	32	39
Age household female	39	35	31	37
Household without male (%)	11	18	14	14
Extended families (%)	18	21	14	19

Source: Mexico survey.

of extended families among tenants in the central city. Equally surprising is that there are so many households without a male among the owners. This phenomenon is especially common in the invasion settlements.

Housing conditions and tenure

Table 4.5 shows that most owners live in an independent house, although some, of course, provide homes for sharers. Most tenants and sharers live in rooms or apartments. Owners have much more space than either tenants or sharers, but their homes are generally flimsier and they have worse infrastructure and services. More owners live in houses with zinc roofs or with an illegal electricity supply than do tenants or sharers. This difference in housing conditions is explained principally by the different ages of the settlements and, in practice, conditions for owners vary considerably between settlements. Conditions for owners are better in

Santo Domingo or El Sol than they are in Avándaro or Belvedere. Ownership offers households the possibility of better living conditions. But, except in terms of space, these are better conditions in the future rather than now. Owners in newly formed settlements have to suffer poor living conditions for several years.

Table 4.5 Housing conditions by tenure, Mexico City

Variable	Owners	Tenants	Sharers	Total
Space available (m²)	39	24	25	3
Households with one room of exclusive use (%)	25	54	48	37
Earth or cement floor (%)	54	39	52	46
Zinc or waste roof (%)	44	18	36	32
Water tap in house (%)	25	45	44	36
Sewerage connection inside house (%)	32	59	56	46
Irregular electricity service (%)	51	2	8	26

Source: Mexico survey.

Preference for ownership

Most families value home ownership. All but nine of the 303 owners said that they preferred to be owners; of these, six said that they were indifferent about the alternatives, and the three others expressed a preference for renting. When asked to elaborate on their preference, only 2 out of 291 owners said that they gained nothing from ownership. The rest cited security of tenure, a sense of independence, cheapness, more space and "having something of one's own". The last point was mentioned by 38% of the respondents, and a sense of independence by a further 27%. Of course, many recognized that ownership also brought certain problems, not least when it came in the form of self-help accommodation. While 54% said that there were no disadvantages, 21% mentioned the cost of services and taxes, and 15% that their quality of life was poor. Not surprisingly, owners in the different settlements gave different weight to the problems. In Santo Domingo, for example, they were most concerned about the cost of services and taxes; in Avándaro, where few services are available, they emphasized the poor quality of life.

The residential trajectories of the respondents strongly supported these replies in so far as very few households had given up ownership. Among the 382 households with a residential history of two or three independent homes in the city, 235 had at some time owned a house. Of this total, only four households had moved out of ownership, two into renting and two into sharing. Among the tenants, only 19 out of 240 had ever owned a house in the city, and of these, five had not given it up voluntarily; they had lost it when they were evicted from the land. As such, only 1 in 20 had willingly given up home ownership at some stage of their residential history. Among the sharers, only 6 out of 80 had, at some stage, owned a house in the city and had disposed of it; in most cases because they were old and wanted to move in with their children.

A general preference for ownership was also expressed by the tenants, among whom 58% wanted to own, with a further 25% saying that there was absolutely no advantage to be gained from renting. And yet, a significant minority of tenants clearly had reservations about the advantages to be gained from ownership. Those with reservations tended to live either in the central city or in El Sol, where 55% said that they preferred to rent. For these tenants, the main advantage of renting was either that it was cheaper or, as many in El Sol rather half-heartedly said, "it was somewhere to live". Tenants in the central city were more positive, especially about the favourable location. The main disadvantages of renting were "not having something of one's own" (accounting for more than one-third of the answers), deficiencies in the condition of the house (14%) and the cost of renting (13%).

Of course, an important factor influencing these replies is the issue of cost. The fact that so many of the tenants expressed a preference for renting compared to the replies in the other cities is no doubt linked to

Table 4.6 Rent–income ratios, Mexico City.

Earnings in minimum salaries	Percentage of households in income band.						
	0-4%	5-9%	10-14%	15-19%	>20%	Total	% of income by income band
<1.0	3	0	0	1	1	5	8.4
1.0-1.9	19	16	22	23	17	87	14.6
2.0-2.9	13	28	18	8	2	69	10.2
3.0-3.9	10	6	4	1	1	22	7.7
4.0-4.9	8	6	1	0	1	16	8.3
5.0+	8	5	4	1	1	19	8.9
Total	62	61	49	34	24	230	10.9

Source: Mexico survey.
Note: The 12 tenants with controlled rents in the central city have been included in the frequency counts. They have been excluded from the calculation of income by income band.

the rather low rents that the tenants were paying. The average rent paid was less than one-quarter of the minimum wage;[1] an income which most holds greatly exceeded (Table 4.6). Overall, rent constituted less than 10% of the average household income.[2] Even when the 12 cases of tenants with controlled rents are excluded, the average rises to only 10.9%. The startling fact is that, in the whole sample, only nine families were paying more than 25% of their income in rent. In Doctores, where the preference for renting was expressed most strongly, 47 of the 77 tenants were paying less than 5% of their income in rent. We will return to the question of rents below.

Our interpretation of these results is that most Mexicans have the option of rental accommodation at relatively low rents. They would like to be owners but they are not prepared to be owners irrespective of the conditions. As a result, many of the more prosperous tenants and sharers do not rush into home ownership. They might want to be owner-occupiers but they wish to avoid the trials and tribulations involved in building in the new periphery.[3] But if some tenants choose not to be owners, it is also true that some owners are forced into ownership. The fact that there are many poor, and several female-headed, households in the new periphery, particularly in the invasion settlement of Belvedere, suggests that for some there is little real choice. Faced by the impossibility of paying even relatively low rents, they take advantage of the chance of ownership when it becomes available. Despite the problems involved in living and building in an unserviced settlement, they have no real alternative.

Poor families are likely to gather in settlements where land is very cheap; there are more families initially in the invasion settlements. Nevertheless, it should be remembered that the cost of peripheral land in Mexican cities tends to be cheaper relative to income than that in many other Latin American cities (Gilbert & Ward 1985). At the median time of purchase, for example, the cost of a plot in Santo Domingo and El Sol was less than two months' minimum salary. Given that most only had to put down a deposit, and many were earning more than the minimum salary, it was not that difficult to acquire a plot.

However, other factors besides cost clearly intervene in the decision-making process. One such factor is suggested by the fact that migrants are much more common among the owners than among either the tenants or the sharers. This is particularly true when new owners are compared with

---

1 Calculated on the basis of 25 working days. If it is calculated on the basis of 20 days then the proportion rises to 29%.
2 The mean was calculated for each household and then the average of all the means computed. For comparison, the average rent:income ratio in Caracas was 25%.
3 Gilbert & Varley (1991) have called this group "persistent tenants" in the case of Guadalajara and Puebla.

tenants in the central city. The difference may simply reflect the advantages of natives over migrants in obtaining decent rental or shared housing. They have better opportunities to share accommodation because their families live in the city and because they can find superior rental accommodation through their better contact systems. Natives are likely to choose ownership only when it suits them, and the data on housing conditions clearly show that home-owning natives occupy better accommodation than migrants, whether it is owned, rented or shared. The locals use their family networks to improve their housing situation, the offer of shared accommodation allows them to wait until a good plot becomes available; family loans allow them to buy better quality plots or finished homes in the periphery. Given fewer alternatives, more migrants are obliged to move into lower quality owner-occupation.[1] However, native–migrant differences are also open to another kind of interpretation, one prompted by the strong correlation that exists between current tenure and tenure of the parents' home. For example, whereas 73% of male owners had lived in owner-occupation with their parents, either in Mexico City or elsewhere, among the tenants the figure was only 38%. Perhaps the desire for ownership is passed on from parents to children?

A different interpretation is also possible. Households with children are more likely to be owners than childless families. Having a child seems to stimulate a reaction among some household heads that they should establish a more permanent housing arrangement. It seems to be connected less to the demand for additional space than to the widespread feeling that families wish to have something to leave to their children. Having the first child seems to be more significant than having subsequent children. Clearly, the motivations underlying the desire for ownership are complicated.

Choosing between different forms of non-ownership

Why do some households choose to rent and others to share? Contrary to our expectations, those who share accommodation are generally happy to do so. Sharing does not seem to be a tenure of last resort, chosen by those without the resources either to rent or to buy. Indeed, 54% of sharers said that there were no disadvantages to this form of tenure. Sharers also seem to live in better accommodation than tenants and have higher per capita incomes than most owners.

The principal benefit from sharing is that the household save money, two out of five sharers mentioning this advantage. The main disadvantages are that the household has little independence (13%) and that the

---

1 See Gilbert & Varley (1991) on this point.

property is not one's own (9%). Perhaps as a result, sharing is not a short-term tenure, the sharers averaged almost three years in the current home.

If there are so many benefits, why do more households not share accommodation? Clearly, some tenants do not wish to do so, 18% saying that they value their independence and a similar proportion stating bluntly that they would not like to share. By far the most commonly cited reason for not sharing, however, is that it is simply not possible; 28% gave this as their reason for not sharing and a further 20% said that they did not have parents in the city.

Clearly, kinship ties are crucial in the decision to share, a factor underlined by the finding that every sharer was being accommodated by a member of the family.[1] Most sharers are the children (63%), siblings (16%) or parents (8%) of the owners. Unlike Chileans, Mexicans do not seem to provide accommodation for friends (see Ch. 6).

Around half of the sharers contribute to the costs of the accom-modation, but generally they do not give very much. If they help financially, it is by contributing to the cost of the services or taxes. However, they are just as likely to help in other ways, for example, by helping with construction of the dwelling; 15% of the sharers had built their own accommodation and 33% had helped construct it together with the owners.

When analyzed together with the socio-economic characteristics of the sharers, it is clear that sharing is a favoured choice among young, newly formed households. Only 3 out of 10 had previously had an independent home. The principal advantage of sharing is that it allows them to live in decent conditions even though they have limited incomes; 41% admitted to sharing because they were earning little. There are certain other advantages notably that the their kin can look after children while they work. Perhaps surprisingly, few said that sharing allowed them to save for their own house and only 1 in 11 had actually looked for their own house. Clearly, sharing has its advantages but is equally clearly only available to those with parents or kin who are able and willing to accommodate the young family. Most sharers are therefore natives of the city and relatively few migrants share accommodation.

Residential movement

Whatever their tenure, the population does not move very frequently. Among households answering this question more than half had only two independent homes (including the current home) and 26% only one. Most

---

1 The only partial exception was a household that was being accommodated by the godparents.

households tend to live for a long time in the same house, almost two out of five respondents having lived for at least ten years in the same house (Table 4.7). Among owners in the consolidated settlements, four out of five had lived in the same house for ten years or more. Owners in the consolidated settlements averaged 14 years in the current house, sharers eight years, and tenants in the central city 17 years. While tenants moved more frequently, 73% of tenants in the central area had been in the same house for at least 10 years.

Table 4.7 Years of residence in the present house, Mexico City.

Settlement	Mean tenure (years)	Percentage living more than 10 years in home
OWNERS		
Total	9	40
Avándaro	4	0
Belvedere	6	1
El Sol	13	78
Santo Domingo	15	83
TENANTS		
Total	7	28
El Sol	3	4
Santo Domingo	3	8
Doctores	17	73
SHARERS		
Total	8	36
El Sol	8	35
Santo Domingo	7	38
Total	9	38

Source: Mexico survey.

When families move house, most travel only short distances. Short moves are especially characteristic of the tenants. In the central city 95% of movers had previously rented accommodation within 5km of their present home, in the consolidated periphery approximately two-thirds. The few who move longer distances tend to leave for the periphery.[1]

Among the owners, the pattern of movement is more complex; both the age of the settlement and the way it was formed influencing the pattern. Table 4.8 shows that in the new settlements, owners have moved quite

1 See Brown (1972), Brown & Conway (1980), Gilbert & Ward (1982), van Lindert (1991) and van Lindert & van Westen (1991) for related work on residential movement in the city.

long distances; the mean distance being 11km to Belvedere and 17km to Avándaro. In the more consolidated settlements, however, the distances were much less, three-quarters of the population of both settlements originating within 10km of their current home. The difference between the two sets of settlements reflects the growing physical size of the Mexican capital. When El Sol and Santo Domingo were formed the urban perimeter was much closer to where most people lived. By the time Avándaro and Belvedere were formed, 20km separated the perimeter from the centre of the city.

Table 4.8 Location of previous dwelling of owners, Mexico City.

Distance (km)	New settlements		Consolidated settlements	
	Avándaro	Belvedere	Santo Domingo	El Sol
0–5	11.2	19.1	59.8	43.6
5–10	4.2	23.6	17.9	32.1
10–15	25.0	38.3	14.9	16.6
15–20	30.5	4.4	4.5	2.5
20–25	13.8	8.8	2.9	5.2
Over 25	15.3	5.8	–	–
Total	100.0	100.0	100.0	100.0

Source: Mexico survey.

However, there is a further difference between the pattern of movement in the different settlements. Settlers in the invasion settlements seem to have moved shorter distances than those in the illegal subdivisions. Whereas more than four-fifths of Belvedere families had moved less than 20km, in Avándaro the figure was only two-fifths. A similar, although less marked pattern was also true of the two consolidated settlements. The difference between the invasion settlements and the illegal subdivisions seems to be that settlers in the former are recruited more by word of mouth. One family brings friends, family and neighbours along with them; to improve the chances of success, invasions need to attract as many participants as possible. The settlers also need help close at hand when establishing their new homes. In the case of the subdivisions, the availability of plots in an illegal subdivision attracts households from a much wider radius. Word-of-mouth reports are less important and households are much less likely to know one another when arriving in the settlement.

Despite these differences, movement to all of the settlements shows one clear feature. The majority of households move within the same zone of the city and few have moved to the survey settlements from the northern area. There are clear signs of migration vectors, the families moving outwards within the southeastern and southwestern quadrants of the city.

Landlords

Most landlords in the self-help settlements have smaller families than other owners, and the head of household tends to be older (Table 4.9). There are relatively more single and retired people in this group.

Landlords are also twice as likely as other groups to be self-employed, but they are less likely to work in construction. They live in larger properties than other owners in the more consolidated settlements, and

Table 4.9 Socio-economic characteristics of landlords and owners, Mexico City.

Variable	Landlords	Owners
Number of persons	5.3	6.0
Number of children	3.3	3.4
Age of household male	46	42
Age of household female	42	39
Extended families (%)	20	18
Household income (×1000 pesos)	674	545
Household per capita income	128	91
Male income (×1000 pesos)	565	379
Male employed in construction (%)	9	17
Colour television (%)	23	11
Car ownership (%)	14	12
Years in current house	14	9

Source: Mexico survey.

have lived there rather longer. They have considerably higher incomes and tend to own more consumer durables. Despite these differences, however, they are clearly drawn from the same social class.

Most landlords operate on a small scale, even in the central city where the effects of inheritance have gradually reduced the level of property concentration. In the consolidated periphery, most landlords accommodate only a few tenant households, the average being 2.2. Indeed, less than a quarter have more than two tenants and only 3 out of 40 reported more than four; seven was the highest number of tenant households recorded.

While three-quarters of peripheral tenants pay rent directly to the landlord, only 3 out of 10 do so in the central city. Many of the latter pay a representative of the landlord. In the periphery, although one in six paid rent to a representative, the recipient usually lived on the premises.

While it is a small-scale operation, only a minority of landlords live with their tenants. In El Sol, only 8% lived on the rental property, although in Santo Domingo the figure was 43%. The difference between the two settlements is explicable in terms of the lack of title deeds among many landlords in Santo Domingo. They live on the property for fear that the tenants might claim the land if they did not. Indeed, many owners say that they do not rent because of this danger; in El Sol the landlords have title deeds.

For most landlords, renting is merely a means of generating a small additional income. Three-quarters of the resident landlords receive less than half of their household income from rent. Few landlords have been renting for very long, two-thirds of the respondents had been letting property for three years or less.

When asked about renting as a "business", the response was variable. One-quarter said that it was a good business, two fifths that it was bad, and the remainder saying it was so-so. There was a considerable difference, however, between the landlords in El Sol and those in Santo Domingo. Whereas three-fifths in the former believed it to be a bad business, only one-fifth of the latter thought it so. The difference in opinion is clearly linked to the levels of rent that can be charged in the two settlements. With its high level of consolidation and its excellent communications, Santo Domingo is more attractive to tenants than El Sol. This difference is reflected in rent levels with tenants in the former paying more than twice the amount paid by tenants in the latter.

But the difference in the replies also reflects a difference between types of landlords. For the minority of landlords who had built accommodation specifically to rent it, there seems to be a considerable amount of disillusionment. These "commercial" landlords are not getting their money back in the way they that they had hoped. Probably as a result, few of the landlords are following a capitalist logic. We are dealing here with "domestic renting", the main motive being to supplement low and falling incomes. The landlords concerned enter and leave the activity according to need; they are not investing in housing mainly to rent it out. Their accommodation frequently changes function. It may be built to accommodate the family and later be let because children have grown up and left home. Alternatively, rental accommodation may be used to put up members of the extended family or *paisanos* when the need arises.

The rationale behind renting also has as much to do with saving for the future as anything else. Landlords are not greatly concerned about the low rents they receive because their main reason for building is to own property. Apart from giving them a certain social status, renting offers a form of saving that they trust. Compared to banks or other forms of financial investment, saving through bricks and mortar is something that they understand.

Landlord–tenant relations

Relations between landlords and tenants differ considerably between the central areas and the periphery. In the central areas, relations are more anonymous and more likely to be based on formal contracts. Most tenants in the periphery know their landlord, whereas most in the centre do not. In the peripheral settlements one-third of the tenants already knew their landlord when they first rented the accommodation, many of the others were recommended to the landlord by friends or kin. Between one-fifth and one-quarter of tenants maintain active social relations with their landlord. Indeed, in the periphery certain landlords seem to develop a kind of family *vecindad*. These differences mean that conflict is much more common in the central areas and there are many more complaints from central tenants about their housing conditions.

How do landlords select their tenants? Nearly all of the 40 landlords said that they had specific selection criteria. The great majority preferred to let to families whom they knew (33%), to tenants who had been recommended to them (20%), to relations (15%) or to *paisanos* from their home area (10%). They also had some distinct dislikes. Few liked families with animals or households with more than two children, more than half mentioning one or the other or both!

Few peripheral landlords issued contracts, indeed, only 6 of the 40 claimed to do so and only two issued contracts properly authorized by a public notary. This was very different to the situation in the central city, where three-quarters of the tenants had a contract and one-half had a contract registered with the authorities. In the periphery most landlords did not even demand much in the way of guarantees. One-quarter wanted both a deposit against damage and a month's rent in advance, but many more (38%) only asked for payment in advance and 28% claimed not to ask for any kind of guarantee. In practice, these requests were not enforced because, although most tenants said that they were supposed to pay in advance, 80% of the tenants admitted that they did not do so.

Most tenants claim to pay on time, only one-fifth admitting to having ever paid their rent late, and another 5% admitting to having done this several times.[1] When asked what they would do if they could not pay the rent, the most common answer was that they would look for other accommodation.

Few of the 240 tenants reported that they had problems with the landlord in the previous 12 months, and of the 37 cases where they had, only 24 had approached the authorities, a lawyer or a politician to help resolve the problem. What perhaps is significant is that in every case but

---

1 Only in the central city did we find six families of tenants who admitted not to pay the rent at all.

one the tenants seeking help lived in the central city. In the consolidated periphery, there was the odd complaint but the tenant did not follow it up. Indeed, there seems to be a major difference in attitude and conditions between renting in the central areas and in the periphery. Whereas three-quarters of the central tenants complained about their accommodation, one in four complained in the periphery.

When asked about the rôle of the law, the difference between the two kinds of area was still more marked. In the central areas almost two-thirds said that the law helped the landlord; in the periphery only 3 out of 10 answered the same way, and nearly one-half did not answer the question. The high level of apparent dissatisfaction in the centre, however, did not derive from the tenants belonging to tenant organizations. Although 11% of tenants belonged in the central areas compared to only 2% in the periphery, membership levels were still very low.

Given the more problematic social relations, it is a little surprising that the typical tenancy in the central areas is so long. Tenancies in the last two homes averaged 17 years and eight years, respectively, with 73% having lived ten years or more in the previous house. Even in the consolidated settlements, however, tenancies averaged three years in the present home and five years in the previous one. For the whole sample, the average tenancy of tenants is seven years in the current home and six years in the previous home.

In the light of these figures it is not wholly surprising that although evictions are not uncommon, most tenants leave for other reasons. Among the tenants who had left a previous rented home, 19% had been evicted and another 12% had left because the owner had sold the property. Most had left for a variety of other reasons.

Conclusions

The evidence from Mexico contains several surprises. While ownership is broadly preferred by most households, many families choose not to take up this option even though they have the resources. They do not choose ownership because they can only afford to own in a peripheral self-help settlement where living conditions are difficult. They also recognize that there are certain benefits to be gained from renting or sharing. Apart from offering better living conditions there are important locational advantages, specifically that the household can live near to their work and/or their friends.

The choice of renting is eased in Mexico City by the fact that rents are low relative to incomes. Some in the central city are paying virtually nothing because of controlled rents; very few anywhere are spending more than 15% of their income. Renting is also an option that is rarely made impossible by bad relations with landlords. While evictions are not uncommon, many tenants live for long periods in the same house, particularly in the central area. Most tenants do not get on too badly with their landlords, especially those living in the consolidated settlements.

Sharing is most common among younger households but is not chosen only by those on the lowest incomes. There are positive advantages to be gained from sharing, which those with families living in the city are happy to take. The only reason why many tenant households do not share is that they have no-one with whom they can share.

In sum, there are few clear distinctions between all owners, tenants and sharers. Tenants are not necessarily poorer than owners, sharers are sometimes better off than tenants. Because of the complexity of the housing market, there is a range of housing options available. Of course, none provide wholly satisfactory accommodation but most families exercise some discretion in their housing choice. In making their choice, however, it is clear that tenure is but one factor in a complicated process of decision-making.

Renting is in relative, but not in absolute, decline. Indeed, in recent years increasing numbers of landlords have begun to rent accommodation in the peripheral settlements. Landlords in the consolidated self-help settlements are more affluent than most other owners but are not making a large sum from the business. Indeed, business is the wrong term for what most landlords do. Renting is a way to supplement their household income while accommodation is vacant; it may be given up when members of the family wish to return to the parental home. This form of "domestic renting" is not something that the Mexican government has ever addressed seriously. Without this form of accommodation, however, the housing situation would be dire.

CHAPTER FIVE

Housing in Santiago

Introduction

During the past three decades, the housing market in Santiago has been strongly affected by radical changes in the direction of Chilean public policy. The results of the Santiago survey cannot be understood without some explanation of national politics since the late 1950s.

The political context

Until 1973, Chile prided itself on being the foremost democracy of South America. The ballot box had determined who would occupy the presidency without interruption since 1932. Power had been fought peaceably by three ideologically committed parties, or alliances, representing the political right, centre and left. During the 1960s, politics became increasingly polarized as the main parties began to compete more vigorously for the rapidly expanding numbers of voters, and as fundamental differences emerged between the parties about the appropriate path for Chilean development. After several years of conservative government, a reformist government was voted into power in 1964. Led by Eduardo Frei its promise of a "Revolution in Liberty" embraced a long-delayed attempt at land reform, a massive housing programme, and the partial nationalization of foreign-owned copper corporations. Such reforms satisfied neither the right or the left, and in the highly competitive election of 1970 an alliance of the left, led by Salvador Allende, narrowly won power. The new government sought a "transition to socialism", a programme including a much more radical agrarian reform and full nationalization of the copper corporations. While its social and distributional record was good, the experiment was certainly not an economic success. As a result of bad management and destablization efforts, organized by pressure groups both from within and outside the country, economic conditions deteriorated rapidly. The rate of inflation rose to over 500% and the foreign debt increased dramatically. In September 1973, the military overthrew the Popular Unity government, Salvador Allende dying in the presidential palace.

The leader of the new military regime, Augusto Pinochet, was to remain in power for 16 years. His government was supported by a highly repressive police state: "In its first few years the dictatorship was relentless in its repression: tens of thousands of Chileans were killed, imprisoned, tortured and exiled, a reign of terror unparalleled in Chile's own past" (Collier 1985). Previous methods of social patronage ceased to be employed; there was no room for populist methods, such as turning a blind eye to the invasion of land. Previous patterns of social development, including housing provision, were to be profoundly affected by the new style of government.

Table 5.1 Presidents of Chile, 1952–date.

Period	President	Party
1952–58	Carlos Ibáñez del Campo	
1958–64	Jorge Alessandri Rodríguez	Popular Alliance
1964–70	Eduardo Frei Montalva	Christian Democrat
1970–73	Salvador Allende Gossens	Popular Unity Alliance
1973–90	Augusto Pinochet Ugarte	Military
1990–	Patricio Aylwin	Multi-party alliance

Gradually, the legitimacy of the Pinochet regime withered, and when the dictator sought to extend the period of his presidency through a referendum held in 1988, he was decisively rejected. His nominee in the subsequent election lost to an alliance of the country's main political parties.[1] He was replaced as president by Patricio Aylwin in March 1990, although he continued as head of the armed forces.

Economic development and the distribution of income

From 1930–73, Chile's economic development relied heavily on public-sector involvement, import-substituting industrialization, and the export of copper. Chile was by no means a poor country, but the development model was failing to maintain a satisfactory rate of economic growth (Table 5.2). In 1970, the Popular Unity government tried to introduce a new socialist path to development, but was toppled after three years in power. The Pinochet regime introduced a wholly distinctive approach, adopting a *laissez-faire* model of development. It freed the exchange rate, opened up the economy to imports, sought to expand non-traditional export products, reversed land reform, deregulated the economy, cut back on government subsidies and privatized most public enterprises.

---

1 Only the parties of the far left did not form part of the alliance.

The model was successful in slowing inflation, led to a rapid increase in exports and attracted large sums of foreign investment. After 1975, when there was a serious recession, the pace of economic growth was not unimpressive. From 1976 until 1981, the economy grew annually at 6.9%. Economic progress was interrupted by a catastrophic slump in 1982 and 1983, after which it continued to grow annually by 5.3%.[1]

Table 5.2 Chile: principal economic indicators, 1960–90.

Year	GDP growth rate	Debt repayment /exports	Inflation	Unemployment in Santiago	Minimum wage
1960–69	4.5	na	27.2	na	na
1970–79	3.1	na	175.2	12.1*	131.5**
1980	6.2	19.3	31.2	11.7	100.0
1981	4.1	38.8	9.5	9.0	115.7
1982	-15.7	49.5	20.7	20.0	117.2
1983	-2.4	38.9	23.6	19.0	94.2
1984	5.7	50.1	23.0	18.5	80.7
1985	2.1	43.5	26.4	17.0	76.4
1986	5.5	37.9	17.4	13.1	73.6
1987	4.9	26.4	21.5	11.9	69.1
1988	7.6	21.7	12.7	10.2	73.9
1989	9.3	19.0	21.4	7.2	79.8
1990	2.0	17.9	29.4	6.6	86.9

Note: Figures for 1990 are preliminary and sometimes for only part of the year. *Figures for 1973–79. **Figure for 1970.
Source: UNECLAC (1984, 1990).

Although, by Latin American standards, the growth rate was reasonable, the development model was profoundly unequal. Unemployment in Santiago rose during the first years of the regime reaching a peak of over 30% at the end of 1982 (Rodríguez 1989). By 1976, 57% of Santiago's population could be classified as poor, although that proportion had fallen to 45% by 1985 (Pollack & Uthoff 1989). Admittedly, the situation was better in the capital than in the rest of the country, for in the late 1980s more than 60% of Chileans were living below the UN's prescribed poverty line, and 80% of the country's people were earning less than they had 20 years earlier.[2] The real value of the minimum wage was reduced and by the end of 1987 was worth only US$38 per month,

1 For the 1981–90 period as a whole the annual growth rate averaged 2.6%.
2 *The Guardian*, 5 December 1989.

well below the level in most neighbouring countries.[1] While government welfare programmes managed to prevent any increase in malnutrition or infant mortality rates, living standards among the unemployed were badly affected by the government's economic and social policies (Graham 1991, Meller 1991).

As a result, the distribution of wealth and income became increasingly unequal. Table 5.3 shows how profoundly the shares of household consumption by different income groups changed under Pinochet. The restructuring of the income distribution was reflected in patterns of residential segregation in the city. Eradication programmes removed low-income settlements from affluent parts of the city (see below) intensifying existing social divides. During the 1980s, the bustle of activity in prosperous shopping areas such as Providencia stood in stark contrast to the poverty apparent in the south and northwest of the city.

Table 5.3 Household consumption by income group in Santiago, 1969, 1978 and 1988.

Income group	1969	1978	1988
Lowest 40%	19.4	14.5	12.6
Middle 40%	36.1	34.5	32.8
Highest 20%	44.5	51.0	54.6
Average monthly household consumption (1988 thousand pesos)	75.5	76.3	76.1

Source: Meller (1991) (original source National Statistical Office *Household budget survey for Santiago*).

Politics and the housing question

Over the years, Chilean governments of all political persuasions have sought to raise the level of home ownership. They have achieved this both through offering incentives to the private real-estate sector and by building large numbers of houses themselves (see Table 5.7 below). Subsidies have been offered to favoured social groups under both sets of programmes. Even the Pinochet administration, with its policy of phasing out most forms of subsidy, decided not to dispense with housing subsidies. Chilean society now has a deep-seated desire for owner-occupation, most families realizing that home-owners are treated very generously, both in the private and the public sector.

Housing and land have also been an important arena for political competition. Public housing construction has been one area of struggle; each administration striving to house more families than its predecessor. Sometimes, too, land has been the focus of competition, with invasions

---

1 *Latin American Economic Report*, 31 January 1989.

being supported, implicitly or explicitly, by the major political parties. At times, the political struggle has been so fierce as to lead to a frenzy of land occupation, such as during the late 1960s and early 1970s, when invasions became the normal way for poor families to gain shelter.

The use of repression allowed the Pinochet administration to break out of this practice. From 1973 until 1989, land invasions and illegal land occupation were simply not permitted; wherever irregular settlement occurred it was removed. Also, many families living in existing squatter settlements were evicted. During the 1970s and 1980s, a series of massive squatter removal programmes were carried out in Santiago (CED 1990).

Under Pinochet, self-help housing no longer provided a route for poor Chileans to gain their own home. The only path to owner-occupation now available was through the generously funded, but limited, public housing programme. While those who were eligible for housing support gained substantial benefits, most were simply ineligible. The residual population was now forced to rent or share accommodation but, since the government did little or nothing to encourage rental housing, the number of rental housing units was static. As a result, most of the poor, and indeed many of the middle class, have resorted to sharing. In 1983, 700,000–800,000 people were estimated to be sharing accommodation, almost one-fifth of the city's total population (Rodríguez 1983, Ogrodnik 1984).

Demographic and physical growth
Despite rapid migration from the countryside after the turn of the century, Santiago has not grown particularly quickly during most of the 20th century. When compared with the rate of migration to most other Latin American capitals, the absolute number of Chileans moving into the city was greatly reduced by the early fall in national fertility rates (Merrick 1986). Even during the peak migration period, the 1950s and 1960s, Santiago's growth was not much over 4% per annum. Table 5.4 shows that since 1970 the city has been growing at 2–3% annually. Migration has diminished in importance and natural increase has become by far the most important component in the city's expansion. There are good reasons to believe that during the 1980s net migration to Santiago was close to zero.

Not surprisingly, Santiago's physical growth has also been slower than that of many other Latin American cities (Fig. 5.1). Indeed, despite the spread of suburbia, Table 5.5 shows that the city's physical area has increased more slowly than its population. Population densities have risen gradually as a result; although Santiago is still not a densely populated city by Latin American standards.[1]

---

1 Mexico City had a population density of 151 persons per hectare in 1980 (see Table 3.3) and Caracas an almost certainly underestimated density of 127 persons per hectare (see Table 7.5).

Table 5.4 Population of Santiago, 1907–90.

Year	Population	Growth rate	% Growth due to migration
1907	333		
1920	507	3.3	
1930	713	3.5	57.4
1940	952	2.9	44.4
1952	1,350	3.0	52.7
1960	1,893	4.3	381.1
1970	2,861	4.2	24.4
1982	3,653	2.1	19.5
	(4,318)	(3.5)	
1992	(5,170)	(1.8)	na

Note: Bracketed figures for metropolitan Santiago.
Source: Necochea & Trivelli (1983), CED (1990); 1992 census.

Table 5.5 Physical growth of Santiago, 1900–85.

Year	Area of city (km²)	Annual growth rate (%)	Population density (persons per hectare)
1900	40		na
1940	110	2.6	87
1952	150	2.5	90
1960	210	4.3	90
1970	302	3.7	95
1982	383	2.0	103
1985	409	2.2	105

Source: CED (1990) and calculations of author.

State intervention in rental housing

In recent years, the relative importance of rental housing has fallen dramatically in Santiago (Table 5.6). Tenant households fell from 57% of the total in 1952 to only 20% in 1982. Indeed, between 1960 and 1982, the absolute number of tenant households decreased from 220,000 to 197,000.

During the 19th century most families in Santiago lived in rental accommodation. Rapid migration during the latter half of the 19th century had encouraged both densification in the central areas and the creation of new housing estates on the periphery. Most middle-class families were

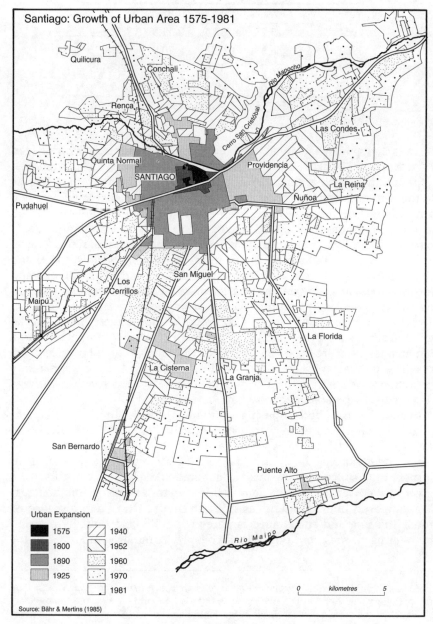

Santiago: Growth of Urban Area 1575-1981

Quilicura
Conchali
Renca
Rio Mapocho
Cerro San Cristobal
Las Condes
Quinta Normal
SANTIAGO
Providencia
La Reina
Pudahuel
Ñuñoa
San Miguel
Los
Cerrillos
Maipú
La Florida
La Cisterna
La Granja
San Bernardo
Puente Alto
Rio Maipo

Urban Expansion

■ 1575		▨ 1940	
1800		◹ 1952	
1890		1960	
1925		1970	
		1981	

0   kilometres   5

Source: Bähr & Mertins (1985)

Figure 5.1

Table 5.6 Housing tenure in Santiago, 1952–82
(percentage of households).

Year	Owners	Tenants	Others	Total
1952	25.7	62.4	11.9	292,649
1960	36.6	52.5	10.9	409,821
1970	57.2	31.0	11.8	638,148
1982	63.6	20.4	16.0	966,889

Source: Soto (1987) for 1970 and 1982; Census 1952 and 1960.

tenants, living increasingly in the monotonous new suburbs, which began to expand after 1880 (de Ramón 1985). Most of the poor were also tenants, although they lived mainly in the central areas of the city, renting rooms in the *conventillos*.[1] This generic term for rental housing actually included a wide range of housing types. At one extreme, *conventillos* were created in the former homes of higher-income groups, the houses being cheaply partitioned and let to poorer families. At the other extreme, new forms of *conventillo* were built on empty plots in the form of badly constructed and serviced *cuartos redondos*. Despite the poor living conditions, the demand for this rental accommodation was considerable; there were few alternative forms of shelter, given the lack of cheap public transport.

In the absence of government controls, landlords were able to make good profits. They could pack large numbers of tenants into these central tenements and were expected to provide little in the way of services. The result was that most families lived in overcrowded and insanitary conditions; one survey of 136 *conventillos* in 1904 recorded an average density of 3.1 persons per room (de Ramón 1985).[2]

Eventually the poor housing conditions, highlighted by frequent epidemics and less frequent protests by the inhabitants, led to a demand for state intervention. The culmination of this political pressure was Chile's first housing law. Law 1838 of 1906 established new councils to inspect and supervise low-income accommodation. The councils were given the power to close or demolish insanitary or unhygienic housing. The law even offered incentives to those prepared to build low-income rental housing and encouraged landlords to build better-quality housing by offering to sell them public land cheaply. A minor component of the

---

1 Although de Ramón (1990) points out that self-help settlements (*guangualíes*) existed on the periphery of the city in the 19th century, founded on land rented from the owners.
2 Not all of the poor, however, lived in the central tenements. Among the very poorest, some were forced to construct their own shacks on marginal land on the edge of the city. The banks of the River Mapocho, the San Miguel Canal and the Aguada Gully were all illegally occupied around this time, although the settlements were removed at times by the authorities.

law also permitted the local authorities to provide public rental housing, the finance to be generated through the issue of municipal bonds.

The councils were effective in making the public more aware of the seriousness of the housing situation and in laying the foundations for later legislation. Their action also led to the closure and demolition of much substandard housing. Between 1906 and 1924, the councils declared 2,216 properties to be uninhabitable and another 1,720 insanitary; they also ordered the demolition of 1,626 *conventillos*, accommodating some 47,000 people (de Ramón 1985). Unfortunately, the law was ineffective in encouraging private investors to build more rental housing. The overall result, therefore, was predictable; rents rose. Rising rents mainly benefited the larger investors whose property, unlike that of many small landlords, had escaped closure. By 1925, rental housing had become a highly profitable business for a much reduced number of landlords; the so-called "age of gold" (Guzmán 1991).

The rental housing boom was undermined, however, by an important rent strike in 1925. The strike was linked to other working-class movements of the 1920s, stimulated principally by rapid inflation. While the strike was short lived, it led to the introduction of rent controls. The government established a housing tribunal, which included represen- tatives of the tenants, and which prescribed the appropriate level of rents. Both the strike and the new controls frightened the larger investors, who began to shift their attention to building accommodation for higher-income groups.

The main physical outcome of the move up-market was the development of the *cité*. This collective form of housing was constructed around a central corridor but, in compliance with the new legislation, was much more solidly built and was provided with better services and communal facilities. Of course, the *cités* were far too expensive for poor families, and were occupied by the middle and upper working classes. Poor families displaced from the *conventillos* moved away from the central city into other forms of rental accommodation. Increasingly, they began to occupy or rent plots of land close to the edge of the city. New self-help settlements were established, built of flimsy materials and lacking most kinds of infrastructure. With the supply of central tenements in decline, this kind of housing option became more common.

Higher-income groups also began to move out to the periphery, although in entirely different directions. Gradually, the city became socially segregated, with the poor living in the south and the rich to the northeast. The central areas also began to deteriorate as the middle classes began to move into suburban accommodation. Investors did not maintain the *cités*, which began increasingly to resemble the decaying *conventillos*.

Despite increasing signs of decay in the central area, the government did little to reverse the trend. The city's Master Plan of 1939 prohibited high-rise development in most of the Santiago Poniente area, so impeding urban renewal beyond the narrowly defined central business district (CBD) (Necochea & Trivelli 1983). The government's overall policy was to encourage families to move out of rental accommodation into home ownership. Rental housing remained a low priority for governments from 1925 until the late 1960s. Only when the tenant associations of the central *conventillos* became active in the highly charged political atmosphere of the late 1960s did rental housing again figure in the political agenda.

Responding to tenant pressure, the Frei government introduced further rent controls and, in their electoral campaign, the Popular Unity alliance sought the tenant vote by promising to fix rents at a maximum of 10% of family income (Lozano 1975). When Allende took over, his government introduced radical changes in housing policy, including stringent controls over rent. "The new maximum rent level was to be computed as 11% of the assessed value of the property for real estate tax purposes. Since the assessed values generally had been considerably lower than the market value, the new rent levels were, in most cases, much lower than before. One result was that many rental houses and apartments were sold to tenants at attractive prices, and cooperatives and condominiums were organised to manage the acquired housing. Also as a consequence of the rent-control law, the speculative construction of housing for middle and upper-income groups . . . came to a virtual standstill" (Lozano 1975).

Government involvement did not always benefit the tenants, especially those who were poor. Certainly, action to improve housing quality in the central area led to widespread demolition. An ambitious urban renewal scheme introduced at the end of the Frei administration, and completed by the Allende government, for example, involved the demolition of a traditional market and a major hospital and led to deteriorated housing being replaced by subsidized high-rise apartments for the upper-middle class (Necochea & Icaza 1990).

The Pinochet government modified the rent-control legislation, although much less than it changed most aspects of housing policy. Law 964 of 1975 allowed tenants to be evicted under a wide range of circumstances, although it prohibited instant eviction and gave most tenants the right to 4–12 months notice. However, the law failed to change the maximum permitted rent. This stayed at 11% of the cadastral value of the property, a figure that remained unchanged in the reforms of 1978 and 1982.[1]

---

1 Law 357 of 2 June 1978 and *Decreto Oficial* 30,146 of 22 August 1978 plus Law 18.101 and *Decreto Oficial* 31.178 of 29 January 1982. As a result of these changes the law established the maximum rent as 11% of the cadastral value plus the cost of the services based on bills and receipts. If a contract was unwritten, the rent was established at the level the tenant said that it was. For eviction or return of the property to be enacted, and where there was no clear leaving date agreed, an eviction order had to be served. This gave four months to the tenant plus two months for every year of the tenant's tenure, up to a maximum of one year. The landlord could demand one month's rent as guarantee against damage.

Nevertheless, rents rose because higher levels of property tax were levied. Since landlords could also evict tenants more easily than before, rents were increasingly determined by market forces. This encouraged more landlords to let property, although their activity was mainly confined to the richer areas of the city. In practice, the rental housing market tended to develop a variety of different forms.

In recent years, many formal-sector landlords in the central areas have held onto their property only in the hope of making capital gains once they could sell the land for redevelopment. Some hoped that the property would deteriorate quickly and be condemned by the authorities, who would then be required to rehouse the tenants. In the meantime, the owners often allowed one of the tenants to sublet the property, thereby taking over the sometimes unsavoury business of collecting rents. In 1985, the rate of deterioration was hastened by the major earthquake that badly damaged the central city. The landlords saw the earthquake as an opportunity through which to regain control over their property, and the state responded through a massive eradication programme. Around 200 ha of the historic central area were cleared of *conventillos*; many families being moved into subsidized owner-occupied housing on the periphery, others simply being evicted. Today, probably only 1 in 20 people in the central city is living in rental tenements.

In addition, changes in the planning regulations in the 1970s removed the stipulation that, within the CBD, buildings must be at least eight storeys high with a continuous facade. Today, few commercial developers build apartments above business premises; they either build single-storey shops or office blocks. The rental housing stock of the centre has been declining fast, and with it the population. Between 1952 and 1982, the population of the Santiago Poniente area fell from 97,000 to 55,000 (Necochea & Trivelli 1983).

Beyond the central area, informal landlords have continued to create homes in the consolidated periphery. Many of these informal landlords are beneficiaries of state subsidies. Given the decline in their real incomes, some poorer families have decided to realize their subsidy by renting out their property. Families who acquired subsidized property in the 1960s and 1970s now have the means to increase their income by letting out rooms. Increasing numbers of families have even begun to rent out part of their plot. The process has been encouraged by the size of the plots (150–180 m^2) and by the difficulty of selling land, given the stringent controls over subdivision. The letting of plots has also been stimulated in recent years by the Catholic Church's Home of Christ programme. The Church makes available prefabricated wooden shacks, 3 m × 6 m, to any family with written permission from the owner to stay on the plot for at least four years. In 1985, around 250,000 of these *mediaguas* had been established throughout the country.

Official encouragement for owner-occupation

Most governments since the early 1940s have encouraged owner-occupation. Every administration has sought to increase the rate of housing construction and most have provided generous incentives for owner–occupiers, in both private and state housing. The importance of state intervention is reflected in the fact that in 1982 26% of the national housing stock had been constructed or subsidized through government schemes (IDU 1990).

The first initiative to boost owner-occupation came in 1943, with the establishment of a Rotating Fund for housing. Financed by a 5% tax on private company profits, the aim was to build homes for working-class families. More significant was the establishment of the Housing Corporation (CORVI) by the Ibáñez government. This action was taken after the results of the 1952 census had revealed the true severity of the housing crisis. Although CORVI then set new records for constructing public housing, the housing deficit continued to grow, particularly among the poor. High building standards priced the accommodation beyond the reach of low-income families; the more affluent were the principal beneficiaries of the programme. Better-off families also gained from the effects of inflation, in so far as CORVI charged too little interest on its loans. Because savers were not compensated for the falling value of their escudos, the housing funds soon became decapitalized.

The failure to build sufficient homes for the poor was reflected in popular demands for a change in government policy. Indeed, during the election campaign of 1957, the capital's first major land invasion took place; the *toma* of La Victoria in September 1957 (see next section). The invasion made housing into a priority for the Alessandri administration. Between 1958 and 1964, public and private construction was expanded and annual production was 2.5 times higher than that under Ibáñez.

The government offered tax incentives to companies building houses with a floor space of less than 140 m^2 and established a new housing finance system in 1959. The National Savings and Loans System (SINAP) permitted financial institutions to link their interest rates to the level of inflation, thereby attracting private savings. The Alessandri administration also offered generous incentives to home buyers.[1]

Public housing construction was greatly boosted and CORVI was given a much increased budget. It received a larger government grant as well as the proceeds from a 5% tax on the profits of private companies. In addition, the social security agencies were compelled to contribute 5% of their funds to CORVI schemes. Between 1959 and 1964, CORVI was able to build 58% of the country's new homes (Trivelli 1987) and in Santiago the

---

1 Law 13305 of 1960.

corporation's huge new apartment blocks accommodated around one-fifth of all the city's families (Lowden, no date).[1]

The huge increase in construction meant that working-class families received public housing for the first time in Chilean history. At the same time, they benefited proportionately less than higher-income groups. Whereas the social security agencies had previously paid workers benefits in line with their total contributions, housing loans were allocated on the basis of ability to repay the mortgage. Under the new scheme middle-class families found themselves in the best position to gain access to the subsidized homes. What was worse was that the poor's contributions to the social security funds were effectively subsidizing the middle-class families.

The Frei administration (1964–70) accelerated housing construction further and promised to build 360,000 units during its 6 years in office. To oversee this massive building programme a new Ministry of Housing and Urbanism (MINVU) was established in 1965 (Cleaves 1974). The private sector, with the aid of state subsidies, would extend its rôle, and two-thirds of total production would be for low-income families, in the form of 50 m² dwellings. CORVI also tried to increase the participation of low-income groups, although nothing was done to alter the workings of the Savings and Loan System.

The goal of building 360,000 homes was soon lowered. Construction was proceeding much too slowly and, in any case, the majority of poor families could not afford to pay even for the smaller houses. The government responded to rising levels of social protest by lowering the standards of housing design. The word "house" was replaced in the official vocabulary by "housing solution" and a huge sites-and-services programme established. *Operación Sitio* was intended as a way of preventing *tomas*, improving housing conditions and increasing family self-reliance. With considerable assistance through the Alliance for Progress programme, the state would provide families with land and infrastructure, the occupants working in groups to build their own homes. The serviced plots of 160 m² were highly subsidized, with settlers paying for the land in instalments. The scheme was launched as part of a major campaign to increase community participation; the political parties, neighbourhood associations and housing committees would all collaborate in the process of housing improvement. Something like 75,000 plots were allocated in Santiago under the *Operación Sitio* programme (Kusnetzoff 1987); 62,739 families being registered on the scheme in one, highly memorable, week (Cleaves 1974).

The housing programme of the Frei administration has been severely

---

1 Previously the funds had only been required to make loans to CORVI; now the money formed part of CORVI's budget.

criticized in the literature, despite the fact that more houses were built than under any previous administration (Table 5.7). By announcing such ambitious housing targets, Frei build up expectations and, by actively encouraging community participation, created the means by which these frustrations could be channelled into social demands. Achievements on the ground could not match these demands and, as the 1970 elections approached, annoyance mounted. Low-income groups began to organize invasions, an action encouraged first by the opposition parties. Soon, the Christian Democrat election machine began to compete by organizing its own *tomas*.[1] While the precise numbers of *tomas* is uncertain, there is no doubt that their rapid growth was linked directly to the approaching election. Kusnetzoff (1975) reports that there were 13 *tomas* in Santiago in 1968, 35 in 1969 and 103 in 1970.

Table 5.7 Houses constructed by administration in Chile, 1953–86.

Period	Public	Private	Total
1953–58	5,949	6,134	12,084
1959–64	17,735	12,730	30,465
1965–70	22,056	17,803	39,859
1970–73	39,089	13,043	52,132
1974–82	5,828	24,051	29,879
1983–86	na	na	42,444

Source: 1953–73 Haramoto (1983) cited in Trivelli (1987); 1974–82, Kusnetzoff (1990); and 1983–86, ICAL (1988) cited in Lowden (no date).

Not surprisingly, the Allende administration (1970–73) gave high priority to housing and announced an Emergency Housing Plan in 1970. It planned to use the housing sector as an employment generator and as a way of accelerating economic growth. The aim was also to use shelter as a means of winning popular support, adequate housing was declared to be "the right of every Chilean family" (Kusnetzoff 1990). The government also responded to popular pressure by ending indexed readjustments to mortgage loan repayments.

The Allende government was no different to most previous administrations, in so far as it was convinced that it was necessary to build finished houses. At first, it was hostile to self-help housing and agreed only reluctantly to complete the *Operación Sitio* programme. It differed from previous governments only in wishing to build many more units and in giving much higher priority to poorer families. It also planned to

---

1 When the Popular Unity coalition had been elected to power, the Christian Democrats stepped up their invasions, even occupying completed housing projects (Kusnetzoff 1975).

channel much more construction through state companies, partly a reflection of its socialist principles and partly an outcome of its difficulties with the private sector.[1] The commitment to building finished homes was eventually to founder on the combination of a lack of resources, poor organization and inadequate technology.

The Allende administration's hostility to self-help housing was based partly on principle. It was also a reaction to the widespread unpopularity of *Operación Sitio*, or what had become jokingly known as Operation Chalk, because of the previous government's failure to provide infrastructure and do much more than mark the outlines of the settlement plots. Eventually, however, pressure from the *campamentos*, and the growing problems afflicting the economy, forced the government to recognize that it was unrealistic to offer good-quality houses to every applicant. Belatedly, it was forced to accept that sites-and-service and squatter-upgrading programmes were inevitable. How its programme would have turned out eventually is uncertain, but in the three years before it was overthrown, it set a record for state-housing construction in Chile (Table 5.7).

When the Pinochet government took over in 1973, its approach to the housing sector matched its broader development strategy. Its main aims were to expand private home ownership in the formal sector, to stimulate private sector construction, and to eliminate state controls over the housing market. In one respect, however, the military regime continued the practice of earlier governments; it offered subsidies to prospective home owners. This approach, in fact, became an increasingly dominant element in Pinochet's housing policy. During the last years of the regime, subsidies for the poor were both numerous and extremely generous.

Initially, the government was preoccupied with illegal settlement and it sought to create a system that would prevent further land invasions. In 1975, it established the "nominee system" which would provide a one-off subsidy for any poor family with some savings and wishing to buy private housing. At the municipal level, to build homes for very poor families, it also set up housing committees, which could either sell the units, with the aid of an interest-free loan, or rent them to those with insufficient resources to pay back a loan. This system of letting accommodation lasted only until 1979, when the Housing Committees were absorbed by the new regional Housing and Urbanization Services (SERVIU) being established in each of the country's 12 provinces in the metropolitan area. The new SERVIUs promptly sold the rental housing to the occupants.

By the late 1970s, fixed subsidies were being offered to poor families wishing to buy homes. Because this generous system was usurped by higher-income groups it was replaced in 1981 by a system of variable

---

1 Relations became especially strained towards the end of the government.

housing subsidies (Kusnetzoff 1990). Subsidies worth three-quarters of the price of the house were offered on the purchase of basic units with a floor space of 25 m^2. The following year, even smaller units became available, so-called "economic houses" of 18 m^2, as well as sanitary facilities, *casetas sanitarias*, of 6 m^2. By 1984, 75% subsidies were extended to those not covered by the previous programmes. Table 5.8 lists the range of subsidies available in 1985. These ranged from core houses through to subsidies on "conventional" private homes.

Although the Pinochet administration followed its principles and shifted most housing production from the public to the private sector, it persisted with the Chilean tradition of subsidizing housing. By the end of 1986, 95,000 families had registered on the SERVIU programme (IDU 1990) and, between 1979 and 1985, 79% of all homes constructed in the metropolitan region were eligible for some kind of state subsidy (Soto 1987). Under an extreme right-wing regime, the state had become the most important source of housing for the majority of families in Santiago; a triumph of pragmatism over principle.

The ideological paradox was still greater in terms of the distribution of the subsidies. The Pinochet subsidies reached large numbers of poor families. Indeed, in the 1979–85 period Soto (1987) calculated that 27% of all forms of subsidy in the metropolitan area went to families in the bottom quintile of the income distribution and a further 28% to families in the next quintile. The most progressive programmes were the *caseta sanitarias*, of which 77% went to the poorest two-fifths of the population, and the "economic houses", of which 62% went to the bottom 40%. Of course, total spending on subsidies was less progressive; Haindl & Weber (1986) estimating that the bottom 30% of income groups received 36% of the subsidies, whereas the richest 40% received 27%.

What helped the Pinochet administration reach so many poorer families was the decision to reduce the average size of houses. Although Soto (1987) describes this as putting "quantity . . . before quality", it can be interpreted more positively as belated recognition that the poor simply could not afford large conventionally built houses. There was also a deliberate move towards a more interventionalist position in the housing market. As Kusnetzoff (1990) puts it ". . . it is quite clear that the government has abandoned its neo-liberal orthodoxy of the seventies. The new accent in its policy is one of increasing subsidization of demand, and particularly in the last two years it has tended to expand its coverage to the poorer sectors of the population."

Even though the Pinochet administration's undoubted efficiency allowed it to get the subsidies to the poor, the housing problem could not be solved in this way. The number of subsidized units was sufficient to house only one-third of the total households formed during those years. In any case, even with such generous subsidies, the broader economic strategy had made too many families far too poor to purchase formal sector homes. Many of those who gained access to subsidized housing were likely to face problems in repaying their loans. Indeed, the level of debt among owner–occupiers in Chile soon reached extraordinarily high levels. By April 1986, SERVIU had 344,641 debtors throughout the country.

Table 5.8 Subsidy schemes available in Chile, 1985.

Scheme	Objective	Conditions
Housing Request System DS 62/1984	Families with low incomes living in marginality withing to acquire a home through SERVIU; the homes are either basic units worth up to 190 UF or social homes of value up to 400 UF	Subsidy: 75% up to a maximum of 180 UF Mortgage of up to 20 years Qualification based on points system based on time, family size, etc.
Sanitation of settlement	Sanitary improvements in zones of extreme poverty, e.g. *Operación Sitio, campamentos*, etc.	Sanitation, basic infrastructure, upgrading and eradication of *campamentos* Maximum value of the sanitation 110 UF
Directed subsidy DS 98/1984	People living in extremely poor urban housing conditions	Maximum subsidy: 180 UF SERVIU mortgage up to 20 years No formal system of approval
Housing subsidy DS 92/1984	Families of modest resources	Maximum value: housing up to 400 UF Subsidy: 165 UF Points system based on savings in system, size of family, cost of plot, etc.
Saving and finance system	Middle-income groups	Scheme A: Credit up to 500 UF, subsidy variable up to 130 UF Scheme B:Credit 500-1000 UF, subsidy variable up to 110 UF Mortgage: fixed interest between 8% and 10%

Note: UF is the Finance Unit. This is index linked. On 12 April 1989 one UF was equivalent of US$18.48.

Source: Ministry of Housing and Urbanism.

And, if the numbers of families with unpaid rents, public service charges and mortgages are aggregated, there were four million people in debt by the middle 1980s (Rodríguez 1989). Poverty had become too deeply entrenched in Chilean society to solve the housing situation through formal sector construction. In the absence of informal sector alternatives, the proportion of families living in shared accommodation rose rapidly.

Land invasions and self-help housing

It will be recalled that the first major *toma* occurred in Santiago during the election campaign of 1957. Two thousand families, whose homes along the Zanjón de la Aguada had been destroyed by fire, invaded land in the La Feria district (Fig. 5.2). The land was intended for a CORVI programme for salaried workers and the invaders were fortunate in so far as CORVI failed to react sufficiently vigorously at the start of the occupation. The invaders also received the influential support of the Primate of Santiago, Cardinal Caro. Once installed as president, Jorge Alessandri was strongly averse to the illegal occupation of land through *tomas*. He eventually turned the La Victoria invasion into a formal self-help programme; a model for the future. He also eradicated a series of small *callampas* ("mushroom settlements"), moving the occupants into new sites-and-services projects, where families lived in provisional housing at the back of the plot and built a proper house at the front.

This early experiment in sites and services was taken up in a big way by the Frei administration. In the *Operación Sitio* programme, every family was to be provided with a plot of 160 m², supplied with electricity, and given access to water through standpipes, although not provided with drainage. It tried to reduce the cost of housing while guaranteeing that the minimum of essential services were available to every family. Unfortunately, by the middle of 1970 "practically every one of the assumptions of *Operación Sitio* had been discarded. The dominating principles of the program were no longer savings over time and self-help. The institutionalization of grievances had given way to conflict resolution through direct action. The concept of gradual improvements of the homesite buckled as well. *Operación Sitio* transformed itself into '*Operación Tiza*' meaning that MINVU simply staked off and assigned lots, and made no attempt to provide them with minimal urban services" (Cleaves 1974). MINVU had been overwhelmed by the flood of invasions and had been forced to join in to help win votes for the Christian Democrats.

As we have already seen, invasions became a fact of life in Santiago between 1968 and 1973 and, by the later date, it is estimated that half a

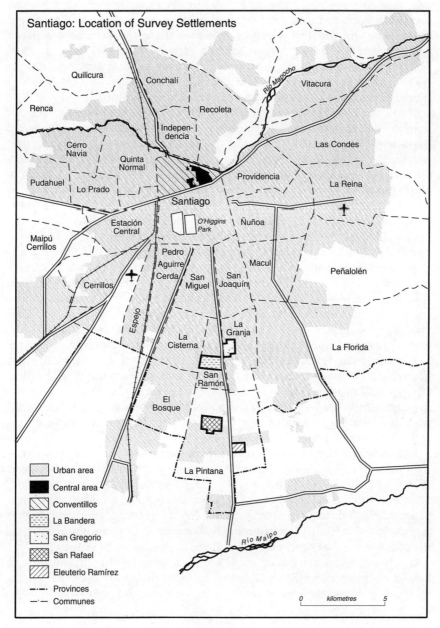

Figure 5.2

million people, 18% of Santiago's population lived in *campamentos* (de Ramón 1990). Nevertheless, it can hardly be said that invasions were willingly encouraged by the authorities under either Frei or Allende. The Frei administration was pushed into supporting invasions as a belated attempt to win the 1970 election, and the Allende government actually disapproved of self-help construction and tried to take over the process of housing construction from the people. However, in comparison with what was to happen under the military regime, the policies of both governments towards self-help settlement were clearly benign.

There can be little doubt that the Pinochet government acted extremely harshly both to new land occupations and to existing invasion settlements. Between 1973 and 1979, the police repressed invasions throughout the country. A series of small *tomas* did take place during the 1980s, culminating in two enormous invasions in 1983, but most of these settlements were removed immediately. Indeed, of the 24 *tomas* that occurred in Santiago between August 1980 and November 1985, all but three were removed instantly (Aldinete et al. 1987). In September 1984, the authorities killed two people in Puente Alto and injured a further 32. The three "successful" invasions were all very distinctive; one occupied church land, in another the private owner came to an agreement with the squatters and the authorities, and in the third the Church was heavily involved. The final case represented a major political confrontation with the authorities, producing two settlements with a combined population of some 8,000 families. Occupying state lands in La Granja, both settlements were named after prominent church leaders, Cardinal Raúl Silva Henríquez and Monseñor Juan Francisco Fresno, in recognition of the help that the Church had provided. Indeed, because of that support the 31,000 squatters were allowed to stay on the land for some time. Gradually, however, the community leaders were eased out and the population relocated into partially urbanized areas nearby, a process that had been completed by 1987 (Kusnetzoff 1990).

Not only did the Pinochet government vigorously oppose new occupations, but it also displaced many existing invasion settlements. Between 1979 and 1985, 28,700 families were moved under MINVU programmes, mainly from land belonging to private owners (CED 1990). The eradication programme led to a major shift in the location of the *campamento* population (Fig. 5.3). Large numbers of *campamentos* were removed from the affluent northeastern part of Santiago to the far south of the city (Fig. 5.4).[1] The relocated families were accommodated either in subsidized homes or, more recently, in sites-and-services programmes. What made this policy especially controversial was that families were

---

1 The major recipients were the communes of La Granja, Puente Alto, Renca, San Bernardo and Pudahuel.

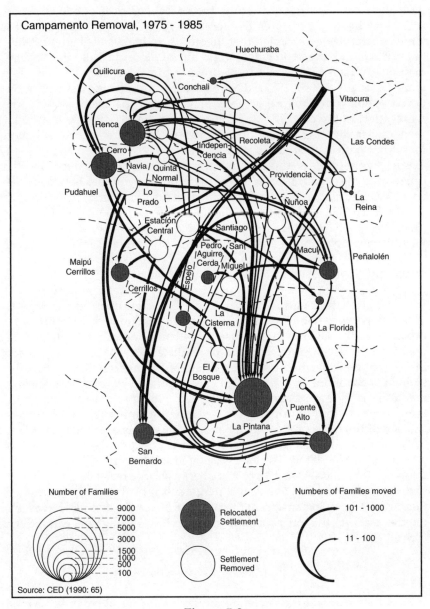

Campamento Removal, 1975 - 1985

Huechuraba

Quilicura

Conchalí

Vitacura

Renca

Cerro

Recoleta

Las Condes

Navia / Quinta Normal

Independencia

Providencia

Pudahuel

Lo Prado

Ñuñoa

La Reina

Estación Central

Santiago

Maipú Cerrillos

Pedro Aguirre Cerda

San Miguel

Macul

Peñalolén

Espejo

Cerrillos

La Cisterna

La Florida

El Bosque

Puente Alto

La Pintana

San Bernardo

Number of Families

9000
7000
5000
3000
1500
1000
500
100

Relocated Settlement

Settlement Removed

Numbers of Families moved

101 - 1000

11 - 100

Source: CED (1990: 65)

Figure 5.3

taken to distant locations, into communes with little capacity for infra-structure provision. The distance factor made the search for work very difficult, a problem accentuated by the deregulation of public transport and huge increases in fares (Figueroa 1990). In a recent survey, most families interviewed accepted that the new housing was of better quality, but 85% found the search for work much more difficult in the new location. In fact, 43% would have preferred to have stayed where they were (Aldinate et al. 1987). The relocation programme created more homogenous communities, accentuating the marked level of residential segregation in the city. The division between the rich and poor areas of modern Santiago is so great that it has been compared to the situation found in South African cities (CED 1990).

Admittedly, the eradication programme was accompanied by the upgrading and legalization of other invasion settlements. Between 1980 and 1987, 139 *campamentos* were provided with services and, between 1979 and 1985, 216,372 land titles were distributed in the in metropolitan region (Soto 1987). 65,800 titles were distributed during 1980, 37,000 in one massive ceremony in the National Stadium (Rodríguez 1989).

The general policy towards illegal land occupation formed part of a wider approach to land in Santiago which Kusnetzoff (1987) describes as conforming to "the dictatorship's strategy for capitalist recomposition". At first, the strategy continued the tradition of government in Santiago to limit urban expansion (Trivelli 1987). The application of strict controls on urban growth sought to increase population densities and thereby maximize the use of services and infrastructure.[1] This policy was reversed in 1979, however, when the urban limits of the metropolitan area were suddenly abolished by Decree 420. The intention was to create a free market in land, reducing state controls and giving total responsibility to the private sector (Rodríguez 1989). Taxes on unimproved land were abolished, the taxes on the buying and selling of lots lowered from 8% to 0.5%, and the state's land reserves, mainly acquired under the Frei government, liquidated. The idea was to reduce the price of building land, a concept that did not work in practice because large real-estate companies had already acquired a good deal of peripheral agricultural land. Although there was a huge expansion in the number of subdivided plots available on the market, the price of land rose rapidly during the early 1980s (Trivelli 1987).

The issue of the allegados

In most Latin American cities, either invasions are permitted or some kind of irregular alternative has emerged through which poor families can

---

1 See Table 5.5 for the effect of this policy.

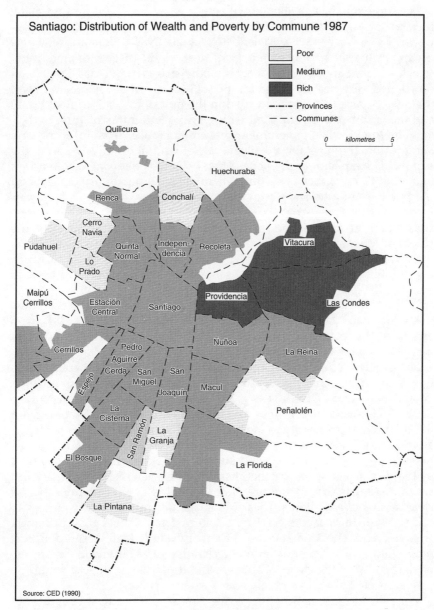

Santiago: Distribution of Wealth and Poverty by Commune 1987

Poor
Medium
Rich
Provinces
Communes

0   kilometres   5

Quilicura
Huechuraba
Renca
Conchalí
Cerro Navia
Pudahuel
Quinta Normal
Independencia
Recoleta
Vitacura
Lo Prado
Maipú
Cerrillos
Estación Central
Santiago
Providencia
Las Condes
Cerrillos
Pedro Aguirre Cerda
Ñuñoa
La Reina
Espejo
San Miguel
San Joaquín
Macul
La Cisterna
San Ramón
La Granja
Peñalolén
El Bosque
La Florida
La Pintana

Source: CED (1990)

Figure 5.4

obtain cheap plots of land on the periphery of the city. In cities such as Bogotá and São Paulo an active business of illegally subdividing areas of agricultural land has developed (Gilbert 1981, Sachs 1990). Around many Mexican cities, illegal subdivisions are now common on *ejido* land (Perló-Cohen 1979, Varley 1985, Mele 1987, Azuela 1989). Plots are relatively cheap, primarily because they lack most kinds of service and infrastructure. In Santiago, the illegal subdivision of land has long been sanctioned very heavily. Land can be sold for urban development only if the owner has a title deed recorded in the official Register of Real Estate and the land has been supplied with services and infrastructure.

When the Pinochet government prevented families from invading land and continued to prohibit the illegal sale of land, it effectively cut off the supply of cheap land to the poor. After 1973, the only way to gain access to land in the periphery was through an official scheme, usually through the purchase of a house. Given the increasing level of poverty of lower-income families, few could afford such homes even with the generous subsidies. The main effect of this policy was the appearance of the *allegados*. Macdonald (unpublished results, 1985) calculated that in 1982 there were 135,000 *allegado* families in the city, while a Catholic Church organization believed that, in the same year, there might have been as many as 190,000 (Bähr & Mertins 1985, Trivelli 1987). Another estimate calculated that 42% of Santiago's nuclear families did not have their own home.[1] Perhaps the most reliable estimate is that of Ogrodnik (1984), who estimated the total to be around 152,000 in 1983, approximatedly 18% of the metropolitan area's total families.

The problem of estimating quite how many *allegados* there are revolves fundamentally around the fact that what constitutes an *allegado* situation is less than clear. Trivelli (1987), for example, follows Macdonald who includes families who are "renting a room, a part of a house, or paying for the right to build a shanty in the backyard of an already occupied site, on a temporary basis". At times the households may combine into a single household in terms of meals and budget; sometimes there is a payment by the *allegado* household, sometimes not. But when is sharing forced and when it is voluntary? Is the presence of adult children in a household a clear sign of *allegado* status? Clearly, some adult children may be forced to remain with the parents for lack of alternatives, others may positively wish to do so. The separation of nuclear or extended families into distinct households really depends on the declaration of the individual members of the family. As such, it is not possible to declare precisely when the condition of sharing is a problem and when it is not.

We will return to the nature of the *allegados* in the next chapter, but earlier work has demonstrated that most *allegado* families are poor and

---

1 *Encuesta Casen*, cited in Necochea (1987). This figure is certainly an overestimate.

many of them are headed by single women. Perhaps the largest group, however, are young families who are in the process of establishing their own home. While most *allegados* are relatively poor, sharing is most common among middle-income families who have room to accommodate an additional family (IDU 1990). Whereas most middle-class *allegados* live in the same house, most poor families occupy a separate shelter on the same plot of land.

CHAPTER SIX

Review of the Santiago survey results

Settlement selection

The settlements surveyed in Santiago are all located between 10km and 16km to the south of the city centre (Fig. 5.1). Because of the military regime's settlement removal programmes, this is one of the city's two most concentrated areas of self-help settlement. One peculiarity of the Santiago survey is that, because new informal housing areas were not permitted after 1973, the "new periphery" could not be represented either by an invasion settlement or by one developed through illegal subdivision. Eleuterio Ramírez, the selected neighbourhood, was founded relatively recently (in 1983) to accommodate families displaced from other areas of the city.[1] As an urbanization designed by the Ministry of Housing and Urbanism and financed by the World Bank, its housing is both formally constructed and fully serviced. The incoming families were subsidized to the tune of 75% of the cost of the accommodation; they were to pay off the balance over a 12-year period.

The settlements in the "consolidated periphery" are more comparable with those chosen in the other cities. Each was built predominantly through some kind of self-help construction. The oldest, San Gregorio, was founded in the late 1950s as part of an official slum eradication programme. Ten thousand families participated in a series of CORVI self-help projects ranging from sites and services through to core houses with 38 m² of floor space; most of the settlement was provided with water and sewerage. The second "consolidated settlement" is La Bandera, a neighbourhood founded by invasion during the Frei administration (1964–70) and then incorporated into the government's *Operación Sitio* programme of *callampa* eradication. The third "consolidated" settlement,

---

1 Mainly from Renca, Conchalí and Pudahuel in the north of the city and from La Cisterna and La Granja in the south.

and the best example of a self-help community, is San Rafael. Established when 2,500 families mounted a land invasion in 1961, the settlement was provided with some services and infrastructure by the Frei administration. Further physical improvements were made under Allende and legal titles were provided by Pinochet in 1980.

The survey of the central tenant population was conducted in 38 *conventillos*; tenants in other kinds of property in the city centre were not interviewed. The *conventillos* are located in an area of some 7km² occupying a zone flanked by the Mapocho River to the north, the Ñuñoa railway station to the east and the San Eugenio railway works to the southwest. The area was badly affected by the 1985 earthquake.

The prohibition on self-help development in Santiago has clearly had a knock-on effect on renting and sharing. It excluded many poor families from ownership and produced the phenomenon of the *allegados*. We have already stressed the importance of the *allegado* problem in Santiago and this was felt to be sufficient justification for conducting 80 additional interviews with these families. The *allegados* are defined as those households living in the same house as another and eating at the same table. Sharers, by contrast, are households occupying the same lot as another family but living in a separate home and eating separate meals.

The nature of owners, tenants, sharers and allegados

Owners have higher household incomes than non-owners and own much more in the way of consumer durables. Fewer male owners are employed and those who have jobs are more likely to be self-employed or to be engaged in the construction industry. Owners are considerably older than most non-owners and, as a result, heads of owner households are much more likely to be women. Owner households tend to be larger than those of non-owners and to contain more children.

Tenants have lower household incomes than owners but because they are much smaller in size, they earn more in per capita terms. They are more likely to have industrial jobs and are less likely to be self-employed. Heads of household are generally younger than adult owners but are older than adults in sharer households. However, tenant households differ greatly between themselves. Small families are much more common in the *conventillos*, where heads of household are older and more likely to be children of tenants.

Sharer households are of similar size to those of tenants but have lower household and per capita incomes. Household heads are younger than those of other tenure groups, except for the *allegados*. Most households contain an adult male and there are few extended families. Most males are in paid work but rather few females are employed.

Families accommodated as *allegados* are usually both small and nuclear in form. The household head is typically young, and many families lack an adult male. Households accommodating *allegados* have higher total incomes than any other tenure group but, because of the large size of the combined family, they are actually poorer. They earn less in per capita terms than other kinds of household and own fewer consumer durables.

Income and tenure

Table 6.1 shows that, even if owners have higher household incomes than tenant or sharer households, they are worse off than tenants in per capita terms, although still better off than the sharers. These overall differences, however, do not mean a great deal because of the variations within tenure groups. Among the owners households are considerably better off in Eleuterio Ramírez and San Rafael than in the other settlements. Among the tenants, those living in the *conventillos* are much more affluent; indeed, the central tenants are the most affluent group in the sample. Within settlements, the tenants in the consolidated periphery have lower household incomes but, because of their smaller household size, are actually better off than the owners. Sharers have lower household incomes than either owners or tenants, although within their own settlements their per capita incomes are about the same as those of the owners. *Allegado* households have high incomes but when their large size is discounted they prove to be the poorest tenure group.

Table 6.1 Household income by tenure and settlement, Santiago (thousand pesos).

Settlement	Household income	Per capita income	Male income
OWNERS	33.3	6.3	22.3
Eleuterio Ramírez	35.1	7.2	20.2
San Rafael	34.5	6.9	25.3
La Bandera	31.9	6.0	21.6
San Gregorio	31.9	5.5	23.1
TENANTS	28.5	7.3	21.8
La Bandera	27.3	6.2	25.2
San Gregorio	28.4	7.3	20.1
Conventillos	29.6	8.5	19.9
SHARERS	22.4	5.7	19.3
La Bandera	18.7	5.1	18.5
San Gregorio	26.2	6.4	20.2
Allegados	38.3	5.0	19.0

Note: The figures for owners exclude landlords and those for *allegados* include the host family.

Source: Santiago survey.

Table 6.2 shows that owners are more affluent than non-owners in terms of their ownership of consumer durables. While there is some variation between settlements, tenants are slightly better off than sharers, and both tenants and sharers are better off than the *allegados*.

Table 6.2 Wealth indicators by tenure, Santiago (percentage with article).

Article	Owner	Tenant	Sharer	Allegado	Total
Radio	85	80	67	54	79
Colour television	28	18	15	9	22
Sewing machine	38	25	9	7	29
Refrigerator	52	27	24	9	9
Car	11	?	0	0	3
Bicycle	43	29	31	12	34

Source: Santiago survey.

Employment and tenure

Because owners are much older, owner households contain many more retired people. There is also a higher proportion of women owners in paid work than among the tenants or the sharers. Conversely, among the younger sharers and *allegados* there is a higher proportion of males in paid work. Among the *allegados*, however, many families lack a male adult and a much higher proportion of women are working.

Table 6.3 Employment by tenure, Santiago.

Variable	Owner	Tenant	Sharer	Allegado
Male employment in industry	17	24	38	29
Male employment in construction	24	18	13	16
Males in paid work (%)	72	82	93	93
Females in paid work (%)	21	17	8	31
Self-employed males (%)	37	29	22	13

Source: Santiago survey.

More owners than non-owners are engaged in the construction sector, although many building workers are also found among the tenants (10%), sharers and *allegados* living in the periphery. Industrial employment is lower among owners and higher among non-owners. Self-employment is much lower among sharers and the *allegados*.

Age and family structure

The Santiago results are very similar to those found in an earlier study of age and tenure in Bogotá, Mexico City and Valencia (Gilbert & Ward 1985). Owners are generally older than non-owners. Male heads of owner households average 48 years of age compared to 37 years among the tenants, 31 years among the sharers and 30 years among the *allegados*. Too much should not be made of these differences, however, for there are considerable variations within the tenure groups. Owners in San Gregorio, for example, average 56 years, whereas owners in Eleuterio Ramírez average only 38 years. Similar variations are apparent among the tenants; in La Bandera heads of household average 33 years, in the central *conventillos*, 43 years.

These variations in age of household head by settlement reflect the peculiarities of the Santiago land and housing markets. The prohibition on self-help housing since 1973 means that few young families are to be found in the new periphery compared to the situation in most other Latin American cities. In addition, because of the city's very low rate of residential mobility (see below), there is a close link between the formation date of the settlement and age of the head of household. Structural factors also influence the average age of tenants. Because of rent control in the central city, many tenants have been living in the *conventillos* for a long time. They are, therefore, much older than tenants in the other settlements.

Table 6.4 Age and family structure by tenure, Santiago.

Variable	Owner	Tenant	Sharer	Allegado
Number of persons	5.3	3.9	3.9	7.7
Number of children	2.4	1.9	1.9	2.4
Size of nuclear family	4.0	3.6	3.9	3.7
Age household male	47.6	37.2	31.1	29.6
Age of household female	47.3	36.2	27.2	28.1
Household without male (%)	27.3	16.8	6.2	37.1
Extended families (%)	43.7	15.9	1.3	92.3

Note: Figures for the *allegados* exclude the host family except for the number of persons and the percentage of extended families.
Source: Santiago survey.

Owner households are generally larger than those of tenants or sharers. Given the advanced age of many of the owners, many families lack a male adult, among owner households in San Rafael the proportion without an adult male reaches 40%. Widowhood is clearly linked to age and this is a significant factor in San Rafael where women heads of household average 47 years and where almost one-quarter are over 60 years of age. By contrast, the younger tenant and sharer households contain a far higher proportion of male heads of household, even though 17% of tenant households lack men; 26% in the *conventillos*.

Extended families are very much more common among the owners than among the tenants and sharers. Again, however, there are important variations due to differences in the age of household heads by settlement. Among the owners, 61% of households in San Gregorio contain extended families, compared to only 22% in Eleuterio Ramírez.

The family structure of the *allegados* is highly distinctive. Some 37% of nuclear households lack a male head and as many as 14% lack a woman. The latter finding is particularly significant as none of the other tenure groups feature more than a handful of families without adult women.

Housing conditions and tenure

Table 6.5 shows that owners have much more space than other tenure groups. Whereas substantial minorities of the tenants and sharers live in one room, virtually all owners have two or more rooms. Neither sharers nor tenants in the *conventillos* have much space. Owners also benefit from better services, although there is not much difference with respect to electricity supply. Table 6.5 suggests that sharers have less access to services than tenants, but the difference is explained largely by the better services available in the central city. If the central tenants are excluded, there is little difference in the services available to the two groups. In terms of the quality of the house, certain households in every tenure group live in accommodation with flimsy roofs and earth floors.

Table 6.5 Housing conditions by tenure, Santiago.

Variable	Owners	Tenants	Sharers
Space available (m²)	38	25	22
Households with one room of exclusive use (%)	3	37	46
Earth or cement floor (%)	11	10	18
Zinc or waste roof (%)	10	21	4
Water tap in house (%)	96	68	68
Sewerage connection inside house (%)	97	72	45
Irregular electricity service (%)	7	8	6

Source: Santiago survey.

Preference for ownership

In Santiago, most tenure groups express a general preference for ownership and, among tenants, 86% say that they would prefer to be owners. Only among the *conventillo* tenants is there a substantial minority (30%) which prefers their current tenure. Among sharers, the proportion preferring ownership is slightly lower than among tenants, but still accounts for 73% of the total. The lowest preference for ownership is found among the *allegados*, 64%. This general preference for ownership does not seem to be influenced greatly by income level; non-owners of all income groups would prefer to be home owners.

Table 6.6 Tenure preferences among non-owners, Santiago.

Preference	Tenant	Sharer	*Allegado*	Total
To share	0.0	22.0	34.7	10.9
To rent	13.8	5.1	1.3	9.7
To own	86.2	73.4	64.0	79.4
Have looked for own home:				
No	60.0	40.0	42.3	51.7
Yes successfully	5.8	3.8	7.0	5.6
Yes unsuccessfully	34.2	56.3	50.7	41.7

Source: Santiago survey.

The strength of these preferences is not reflected, however, in the intensity of the search for owner-occupation. Despite their general preference for ownership, as many as 60% of tenants, 40% of the sharers and 42% of the *allegados* had never looked for their own home (Table 6.6). In the central area 78% of the tenants had never looked. Income levels certainly do not explain why more sharers and *allegados* looked than tenants, although they do explain the chances of success among those who did look for homes. Whereas 60% of those with monthly incomes higher than 60,000 pesos had been successful, only 15% of those earning less than this figure had found a home.

What these figures suggest is that tenants are generally less happy with their current tenure and are therefore keen to move into home ownership. Many, however, realize that they may not be able to do this in the peculiar housing market of Santiago. Many of those who have not looked for their own home are ineligible for subsidized ownership. Central tenants are also discouraged by the realization that ownership is only feasible if they are prepared to move to the distant suburbs.

That so many households should prefer ownership is entirely rational in a context where the state has strongly subsidized owner-occupation. The fact that many owners benefited from uncollected debts on their state housing has made ownership still more desirable; a preference accentuated when the state declared a moratorium on debt collection in 1989. Tenants have received nothing comparable in the way of benefits from the state and, consequently, many tenants view this tenure as a means of waiting for ownership.

It is also clear that rents are high as a proportion of income. The average rent in the survey was 5,042 pesos, 30% of the minimum wage at the time. On average, tenant households paid 34% of their income in rent. Even allowing for some under-recording of incomes at the bottom of the income distribution, this is an extremely high proportion in comparison with the figures for Caracas and Mexico City, and certainly when measured against the international "norm" of 20% used in many housing projects. Among those earning less than the minimum salary, a substantial proportion of the sample, the share was 65%; clearly a level unsustainable over anything but the very short term. Only above two minimum incomes does the rent burden become tolerable. In the light of these figures the aversion to renting is fully justified.

Table 6.7 Rent:income ratios, Santiago.

Earnings in minimum salaries	Number of households by percentage of income paid in rent						% of income by income band
	0-9%	10-19%	20-29%	30-39%	>40%	Total	
<1.0	6	17	7	9	33	72	65
1.0-1.9	19	27	18	10	5	79	20
2.0-2.9	16	18	8	1	1	44	15
3.0-3.9	2	6	2	0	0	10	15
4.0-4.9	6	2	1	1	1	11	17
5.0+	5	1	0	0	0	6	7
Total	54	61	36	21	40	212	34

Source: Santiago survey.

However, the benefits from ownership are not only financial. When asked about the major benefits to be derived from ownership, 77% of the owners stated that ownership gives them greater security and "peace of mind" (*"por la tranquilidad"*). The second most frequently mentioned reason for preferring ownership is to avoid paying high rents, although this was only mentioned by 7% of the owners. Housing conditions were rarely mentioned, but it will be recalled that owners had access to more space and better services than tenants or sharers (Table 6.5).

This general preference for ownership is supported by the residential trajectories of most of the interviewees. Among the owners in every settlement, a majority (70%) had rented their previous home and 23%

previously shared. Not surprisingly, few among the non-owners had previously been owners. Among the tenants 81% were previously tenants and 12% had previously shared.[1] Among the few sharers and *allegados* with a previous independent home, the vast majority had previously rented or shared.

Preferences, however, seem to have little to do with past experience. Certainly, there is little discernible difference between owners and tenants in this respect. The two have virtually identical proportions of parents owning their own home. There is no tendency for owners to be drawn disproportionately from the children of owners, the only possible exception being that the population in the *conventillos* seem to be predominantly the children of tenants.[2] Nor is there any tendency for migration to affect the tenure choice between ownership and renting. Just over half of male owners and tenants had been born in the metropolitan area. Where migration did affect tenure was in reducing the numbers of migrants in the sharer and *allegado* populations. It is not possible to share a home if you do not have kin in the city.

Choosing between different forms of non-ownership

In contrast to the situation in Caracas and Mexico City, a much lower proportion of the population rents accommodation in Santiago. One explanation may be that rent levels are relatively higher in Santiago. The rents do not seem commensurate with the quality of the accommodation on offer. Perhaps for this reason, a considerable proportion of the non-owning households seems to have taken some kind of sharing option. Certainly, few sharers seem to be very keen to rent. Whereas 73% would prefer to buy and 22% are happy to share, only 5% said that they would prefer to rent. Among the *allegados*, the preference for ownership is slightly less marked, and the preference for sharing higher, but only one family declared that they would prefer to rent.

If sharers and *allegados* do not want to rent, why do current tenants not share accommodation? The most common response to this question is that it is not possible because of lack of space in the parental or family home. If we add to this the lack of parents in the city and the reluctance of the parents to accommodate the household, then the inability to share accounts for 36% of the replies. Equally important, however, is the fact

---

1 The fact that 7% of tenants said that they were previously owners may well be explained by their having lived with parents who had been owners.
2 Presumably some of the tenants have inherited the tenure of a room in a *conventillo* from their parents.

that 42% of the tenants have either not thought seriously about this possibility or do not want to share.

Further factors underlying the choice between renting and sharing are clearly linked to household circumstances. It will be recalled that tenant households differ greatly from one another, in particular, that central tenants are very different from those living in the periphery. Tenants living in the central city are much less likely to have looked for their own home and express themselves to be much happier with renting than other tenants. It is also surely not incidental that tenants generally, and especially those living in the central city, are more affluent than the sharers and *allegados*.

There are also some seemingly telling differences in the household structures of tenants, sharers and *allegados*. The last group contains a considerable proportion of households without a married couple: 31% of households lack a male head and as many as 16% lack a woman. The proportion of married couples varies from 79% among the tenants, to 91% among the sharers to only 54% among the *allegados*. The last category seems, therefore, to contain more households in difficult circumstances; the *allegado* population seems to be the group most clearly forced into their current housing circumstances.

Residential movement

In recent years, at least, Santiago seems to be characterized by a very low level of residential mobility. Once a low-income family acquires a home, it does not seem to move. In the consolidated periphery the median length of residence in the current home is 23 years, most of the families having lived there since they first became home owners. In San Gregorio, the mean is 27 years and in San Rafael 21 years.

A high level of stability is also quite common among the tenants, particularly among the central tenants, who average 12 years in the current home (median 8 years). Of these, 46% have lived more than 10 years in the current home and 18% more than 20 years. Of central tenants, 74% have had no previous rented home, and 25% of *conventillo* tenants have lived more than 40 years in their current home. By contrast, the average stay of tenants in the periphery is much shorter. However, the average tenure of three years is hardly a sign of rapid movement; 55% of tenants are living in their first home and only 14% of tenants have lived in more than two rented homes in the past five years.

As a result of these long tenures, comparatively few households have had a previous independent home. Only 32% of owners declared that they had a previous independent home in the city, 55% of tenants, 15% of sharers and 24% of the *allegados*. Families in all tenure groups seem mainly to have grown in their existing house. Having children seems to

Table 6.8 Years resident in the current house, Santiago.

Settlement	Mean tenure (years)	Percentage living more than 10 years in home
Owners		
Total	17.9	24.0
Eleuterio Ramírez	5.9	1.3
San Rafael	21.4	77.0
La Bandera	17.3	89.1
San Gregorio	26.8	91.1
Tenants		
Total	6.1	17.9
La Bandera	3.0	2.6
San Gregorio	3.6	4.9
Conventillos	11.6	46.0
Sharers		
Total	6.3	16.3
La Bandera	5.4	12.5
San Gregorio	7.2	20.0
Allegados		
Total	6.7	23.6
La Bandera	5.6	18.0
San Gregorio	7.8	28.2

Source: Santiago survey.

have had little effect on mobility and not to have triggered a house move, a common characteristic of house moves in most developed countries (Doling 1976, Clark & Onaka 1983).

Landlords

In terms of their household characteristics, most landlords seem to be very similar to other owners in the same settlement (Table 6.9). They are comparable in terms of age, have similar numbers of children, do similar kinds of work, and have lived equally long in the same house. While they have lower household incomes, their per capita incomes are very similar.

However, there are a few interesting differences. First, landlords tend to have smaller households than other owners, on average containing one person less. Secondly, a very high proportion of landlord households lack an adult male or an adult female. In La Bandera and San Gregorio, 38%

Table 6.9 Socio-economic characteristics of landlords and owners, Santiago.

Variable	Landlords	Owners
Number of persons	4.3	5.3
Number of children	1.9	2.4
Age of household male	51.1	47.6
Age of household female	52.3	47.3
Extended families (%)	35.7	43.7
Household income (pesos)	28,024	33,336
Household per capita income	6,517	6,290
Male income ($\times 1,000$ pesos)	22,105	22,315
Male employed in construction (%)	39	24
Irregular electricity supply (%)	7	7
Colour television (%)	31	28
Car ownership (%)	10	11
Years in current house	23	18

Source: Santiago survey.

of landlord households lack a man and 14% a woman, compared to figures of 25% and 3% among other owners in those settlements. Whereas 72% of the owner households are headed by married couples, only 48% of landlord households fall into this category. Thirdly, a higher proportion of landlords work in the construction industry.

This description suggests that some owners have become landlords because of the difficulty of combining their work with the caring of offspring. This interpretation is supported by the distribution of income among the landlords. While some are clearly among the more affluent of their communities, one-quarter of landlords in La Bandera and two-fifths of those in San Gregorio are among the poorest.

Many landlords, therefore, are letting out property as a kind of subsistence strategy. A considerable number, for example, are effectively renting out part of their land, with the tenants building a rudimentary house in the backyard; a practice encouraged by the Church's *Hogar de Cristo* (Home of Christ) programme (see previous chapter). None of the landlords rent furnished accommodation.

Among the 42 landlords interviewed in the consolidated periphery, 30 had only one tenant and a further nine only two. No landlord had more than five tenants. Many were unimpressed with the money they receive from letting; 40% saying that renting is a bad business, and a further one-third saying that it is "so-so". The average landlord obtains one-quarter of his or her income from rent. This is quite high given the kind of property being rented out, but is clearly not the basis of a commercial enterprise. However, it should be remembered that there is some diversity among the landlords. While some are poor, others are obtaining a decent income from their accommodation. A majority of landlords have only been letting rooms or space for a short period. Some 35% of landlords have been doing it for less than three years, although 26% have been renting out accommodation for more than seven years.

Landlord–tenant relations

Few of the tenants have had many rented homes and, to judge from their replies, there is seemingly little pressure on tenants to move house. One in four respondents mentioned that they had been evicted, and a further one-tenth that they had moved because the rent had been raised. However, one in five said that they had moved to obtain better accommodation, and a similar proportion reported that there had been a number of factors underlying their move.

When asked about their current landlord, only 10% of tenants admitted that they were having problems. Many more did comment unfavourably about the quality of their accommodation, although the nature of the problems differed between the settlements. In the *conventillos*, and to a lesser extent in San Gregorio, the main problem lay in the deficiencies in the services. However, in another consolidated settlement, La Bandera, tenants were more concerned about the lack of space.

The landlords were readier to admit to having problems with their tenants; 22% said that they had a problem in the previous 12 months – drunkenness being the most frequently mentioned complaint.[1] However, if their use of the courts is any criterion, the problems have not been serious; only 4 out of 42 landlords had used the legal system.

Perhaps the fact that complaints are not higher is linked to the way that landlords select tenants. Most tend to rent only to families they know or who have been recommended to them. Although one in five landlords say that there are no specific criteria influencing selection, only 3% say that they will accept any kind of family. The specific grounds on which landlords reject applicants are a little vague, the majority simply stating that they do not accept families if they "inspire a lack of trust"! Too many children may possibly be sufficient cause for such a lack of confidence, although only one in eight landlords mentioned this criterion specifically.

Neither the law nor interest groups seem very influential in landlord–tenant relations. Landlords did not seem to think that the rent laws had any effect on landlord–tenant relations and they seem remarkably vague about the nature of recent legislation. Few landlords

---

1 Perhaps landlords are more likely to complain about tenants because each had more tenants than tenants had landlords.

issue contracts and few landlords seem to demand more than one month's rent in advance as a guarantee. Nor do they belong to landlord associations or use rental administrators. In contrast to the landlords, one in five tenants belong to some kind of housing association. However, few of these associations play any confrontational rôle with the landlords, especially those in the periphery.

Conclusions

The housing market in Santiago is very different from that in most Latin American cities. It is distinctive in so far as there has been no illegal land development on the periphery of the city since 1973. As a result, the new self-help settlements, which proliferate elsewhere, are absent in Santiago. This prohibition on self-help development on peripheral land has affected the rest of the housing market. It has pushed up rents to levels higher than those in Caracas or Mexico City and, since only the households that can pay the high rents remain as tenants, the per capita incomes of tenant families are higher than those of other tenure groups. Poorer households, unable to pay the high rents, have been forced out. The latter have been unable to gain access to cheap forms of owner-occupation and have therefore moved into other kinds of tenure situation, becoming sharers and *allegados*.

Home owners are distinctive in Santiago because they have either acquired property at times when land invasions were permitted or have gained access to highly subsidized conventional accommodation. Few newly formed poor households have become owner–occupiers in Santiago in recent years, and poor self-help home owners tend to be much older, confined to the more consolidated settlements which they entered many years earlier. Under these circumstances, it is not surprising that owner–occupiers live in better-quality homes than other tenure groups; a major difference from the situation in Caracas and Mexico City.

Rental housing is in decline largely because so few of the population can afford to pay the high rents. Most of the Santiago sharers and *allegados* would, in other cities, be found in rental housing. Such accommodation is in decline in Santiago because many landlords feel that they are not obtaining a reasonable income from their property. There is consequently little effort to increase the amount of rental accommodation.

A further peculiarity of the Santiago housing market is the very low rate of residential movement. Home owners hardly ever move house, sharers and *allegados* remain for many years in the same accommodation and even tenants do not move frequently. The housing market is extremely static.

CHAPTER SEVEN

Housing in Caracas

Introduction

Four factors are critical in understanding the housing issue in Caracas: Venezuela's relative wealth, derived from its position as a major oil exporter and its highly concentrated income distribution; the way that the state has channelled much of its oil revenues into the economy through investment in public building and infrastructure schemes; the political history of Venezuela, and particularly the populist manner in which the main parties have used land, housing and servicing in the highly competitive electoral situation since 1958; and, finally, the difficult terrain, which has limited the amount of available building land and served to encourage rampant speculation. Together these four factors have produced a highly distorted housing system; a combination of luxury homes and high-rise apartment blocks occupying the bottoms of the main river valley while extensive shanty towns cover most of the neighbouring mountainsides (Fig. 7.1).

Income per capita and its distribution

By Latin American standards, Venezuela is a distinctly wealthy country. In 1988, its per capita income was roughly twice as high as that of Chile and 85% higher than that of Mexico (World Bank 1989). Petroleum is almost the sole explanation of this affluence; apart from iron ore, Venezuela exports very little else. Petroleum has been a significant source of export revenues since the 1920s, and the main source of government income since the 1940s when the transnational oil companies were finally compelled to hand over a higher share of their profits. As Table 7.1 shows, government revenues rose significantly at that time and still more dramatically in 1973 as a result of the OPEC price rises. While the 1980s posed difficult economic problems for the country, and per capita income declined by 15% between 1981 and 1988, the impact of the debt crisis and the world recession on the Venezuelan population was greatly cushioned

96

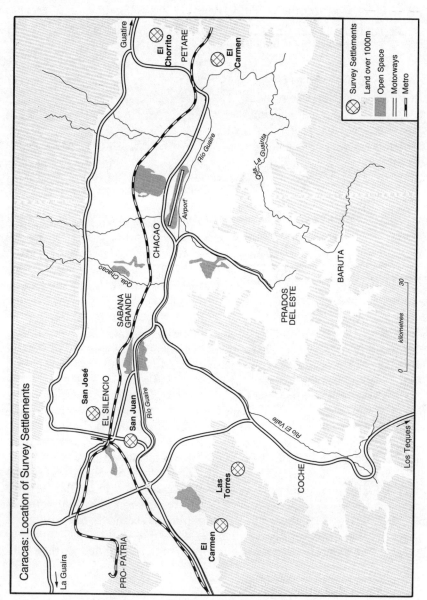

Caracas: Location of Survey Settlements

Survey Settlements legend:
- Survey Settlements
- Land over 1000m
- Open Space
- Motorways
- Metro

Figure 7.1

Table 7.1 Oil exports and government revenues by administration since 1917.

Period	Government	Annual barrels (million)	Annual income (million Bs.)
1917–35	Gen. Juan Vicente Gómez	60	25
1936–40	Gen. Eleazar López Contreras	183	94
1941–45	Gen. Isaias Medina Angarita	227	194
1946–48	Acción Democrática	438	779
1949–58	Military junta and Marcos Pérez Jiménez	730	1,729
1959–63	Rómulo Betancourt	1,094	3,257
1964–68	Raúl Leoni (AD)	1,270	5,114
1969–73	Rafael Caldera (COPEI)	1,274	7,390
1974–78	Carlos Andrés Pérez (AD)	794	30,267
1979–83	Luis Herrera Campíns (COPEI)	636	47,872
1984–88	Jaime Lusinchi (AD)	548	57,946
1989–	Carlos Andrés Pérez (AD)	*	*

*Sources:*1917–73, Fuad (1974); 1974–78 and 1979–83, the author's calculations from Silva Michelena (1987); and 1984–88, BCV (1987) and BCV (*Boletín Mensual* Oct. 1988) – data for oil revenues are for 1984–87.

until 1989 (Gilbert 1989). The severe austerity package that was finally introduced that year savagely reduced living standards and led to riots in the streets of Caracas and other major cities.[1] In 1989, the gross domestic product fell by 8% and prices rose by 84% (Table 7.2).

Economic decline after 1979 led to a considerable fall in per capita income. This decline was superimposed upon a distribution of income which had long been very unequal. As a result, Michelena (1988) estimated that in 1987 between 40 and 50% of all urban families were living in poverty and 10% were living in critical poverty. Palacios et al. (1989) detected a deterioration in the distribution of income between 1984 and 1987, a rise in critical poverty from 16% to 28% and in relative poverty from 29% to 34%. While every income group suffered, the poor suffered most because of a higher rate of unemployment and because of their higher expenditure on foodstuffs. As a result of the 1989 austerity package, the poorest quarter of the population suffered a further 38%

---

1 27 February 1989 will long be remembered as the first day of a wave of violence that led to widespread looting of shops and subsequent police retribution. Officially, 287 people died in the unrest but few doubt that the figure was very much higher (SIC, 513). There have been numerous protests about rising prices and employment in the past 3 years, including a general strike and an attempted coup in February 1992.

decline in income between 1988 and 1989 (España & González 1990).[1]

Despite these depressing figures, it is probable that the number of poor in Caracas living in absolute poverty was lower at the time of the survey than in most other Latin American cities.[2] It is equally clear that, within Caracas, the poor were highly segregated from the wealthy (Fig. 7.2).

Oil and the construction sector

The economic recession hit the construction industry particularly hard. In 1984, the industry declined by one-third and total output was lower than it had been in 1971 (Table 7.2). Unemployment reached 30%, a rise from only 11% 3 years earlier (BCV 1984, 1988).

The recession constituted a major problem in an economy which has traditionally channelled a considerable proportion of its petroleum revenues through the construction industry. In 1978, at the peak of the OPEC oil-price boom, construction accounted for almost 8% of gross national product; by contrast its contribution in 1985 was only 3.3%.

Table 7.3 shows that these fluctuations have had a major impact on formal-sector housing programmes. During the 1980s both private and public housing programmes were badly affected.

Oil revenues, channelled through the public sector, have financed the bulk of construction in Venezuela since 1948 (CEU & OESE 1977). However, the private sector has benefited in a number of ways. First, most public contracts have been contracted out to the private sector construction industry. Secondly, most public building programmes have been constructed on private land and, with little effort over the years to control speculation, this has created many private fortunes.[3] Thirdly, because the state has paid for most forms of urban infrastructure and services, as well as providing most of the cheaper formal housing, the private sector has been able to engage in the more profitable business of building luxury homes. Finally, the private sector has benefited from the industrial activity which the construction programme has generated; steel is the only major building material that the private sector does not produce.

---

1 As a result of the 1989 package, it is estimated that 30% of the population were living in acute poverty; an estimate made by Enrique Rodríguez of the Instituto Venezolano de Planificación and reported by Wilmer Ferrer in "En 30% la pobreza crítica", *El Nacional* 12 September 1989. The fall in incomes, however, came after the questionnaire survey was conducted.

2 It was certainly lower than in Mexico City and Santiago at the time of the survey in 1989.

3 The Lander Report discussed the problems that rising land prices were creating for the housing sector, but its main recommendations were ignored. For a discussion see CEU & OESE (1977) and Negrón (1982).

Table 7.2 Venezuela: principal economic indicators, 1960–90.

Year	GDP growth rate	Construction growth rate	Construction /GDP	Petroleum exports (US $ m)	Inflation	Unem-ployment
1960–69	5.1	1.8	4.5	2,380	1.5	na
1970–79	4.3	9.8	5.5	7,219	8.4	na
1980	-1.5	-16.5	5.8	18,301	19.2	na
1981	0.4	-2.8	6.0	19,094	10.9	6.7
1982	0.7	-8.4	5.4	15,659	7.8	7.7
1983	-4.8	-13.5	5.0	13,667	11.3	11.3
1984	-1.4	-34.4	4.6	14,794	15.7	14.7
1985	0.3	-4.1	3.3	13,144	9.2	13.3
1986	6.8	9.8	4.4	7,592	13.1	11.3
1987	3.0	4.3	6.0	9,054	38.?	9.1
1988	5.8	7.9	6.1	8,023	32.6	6.9
1989	-8.6	-27.1	4.9	9,862	84.3	9.6
1990	5.3	6.7	4.9	13,780	36.1	9.9

Sources: GNP growth: 1960–69 calculated from BCV (1978); 1970–79 calculated from BCV (1978, 1982); 1980, BCV (1982); 1981, BCV (1983); 1982 and 1983, BCV (1984); 1984, BCV (1986); 1985, BCV (1987); 1986 and 1987, BCV (1988); 1988, BCV (1990); 1989 and 1990, BCV (1991).
Construction growth: 1960–69 calculated on the basis of BCV (1978); 1970–79 calculated from BCV (1978, 1982); 1980, BCV (1982); 1981, BCV (1983); 1982 and 1983, BCV (1984); 1984, BCV (1986); 1985, BCV (1987); 1986 and 1987, BCV (1988); 1988 and 1989, BCV (1990); 1990, BCV (1991).
GNP and construction: 1960–69 calculated from BCV (1978); 1970–76, BCV (1978); 1977–78, BCV (1981); 1979–82, BCV (1984); 1983, BCV (1986); 1984–86, BCV (1987); 1987–89, BCV (1990); 1990, BCV (1991).
Petroleum export figures: 1970–76, BCV (1978); 1978–82, Martz & Myers (1987); 1983, BCV (1986); 1984, BCV (1987); 1985–87, BCV (1988); 1988–89, BCV (1990); 1990, BCV (1991).
National consumer price index, December to December: 1962–69, BCV (1978); figures for Caracas, 1970–79 calculated from BCV (1978, 1982); 1981, BCV (1983); 1982 and 1983, BCV (1984); 1984–85, BCV (1987); 1986, BCV (1988); 1987, BCV (1989); 1988 and 1989, BCV (1990); 1990, BCV (1991).
Urban unemployment: 1980, BCV (1983); 1981 and 1982, BCV (1984); 1983, BCV (1986); 1984, BCV (1987); 1985, BCV (1988); 1986, BCV (1989); 1987, BCV (1990); 1988–90, BCV (1991).

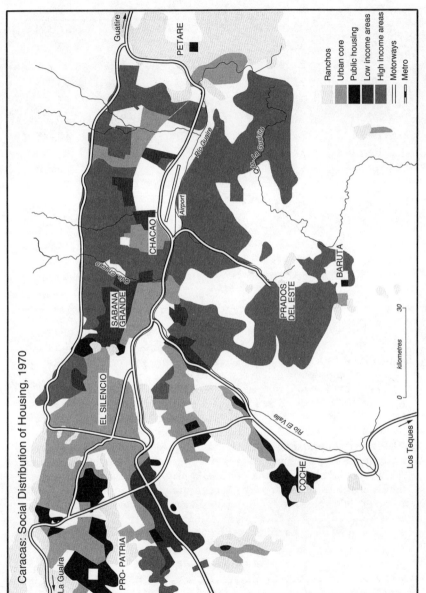

Caracas: Social Distribution of Housing, 1970

Ranchos
Urban core
Public housing
Low income areas
High income areas
Motorways
Metro

Guatire

PETARE

Río Guaire

Airport

CHACAO

Qda. Caracas

Qda. La Guairita

SABANA
GRANDE

EL SILENCIO

BARUTA

PRADOS
DEL ESTE

PRO-PATRIA

La Guaira

COCHE

Río El Valle

Los Teques

0   kilometres   30

Figure 7.2

Table 7.3 Housing units constructed by public and
private sectors in Venezuela, 1960–90.

Year	Public sector	Private sector	Public/ private	Total
1960–72	242,013	166,408	1.45	408,421
1976	29,346	36,205	0.81	65,551
1977	22,200	41,068	0.54	63,268
1978	23,203	49,975	0.46	73,178
1979	31,830	50,427	0.63	82,257
1980	32,243	51,012	0.63	83,255
1981	34,263	48,552	0.71	82,815
1982	37,137	56,293	0.66	93,430
1983	24,015	28,574	0.84	52,589
1984	15,454	21,682	0.71	37,136
1985	16,558	21,817	0.76	38,375
1986	45,782	23,713	1.93	69,495
1987	42,134	30,169	1.40	72,303
1988	45,300	25,708	1.76	71,008
1989	12,934	16,329	0.79	29,263
1990	28,666	15,631	1.83	44,297

Source: Private construction figures for 1960–72 are from Rodríguez (1976); for 1976–86 from OCEI (1987); for 1987–89 from OCEI (1990); and 1990 from CNV (1991). The original data were provided by FUNDACONSTRUCCION. The figures for the private sector are for licences as opposed to actual construction.

Public construction figures for 1960–72 are from Rodríguez (1976); those for 1976–88 are based on INAVI data, and for 1989–90 from CNV (1991). The figures for the public sector until 1972 are for INAVI only, between 1976 and 1988 they include all public-sector activities except credits, *barrio* upgrading and sites and services. For 1989 and 1990 these additional forms of housing solution are included in the figures.

The total is the sum of the public and private sector totals. Therefore, they sometimes differ from the specific sources listed above. It should be noted that the sources are highly inconsistent and frequently differ from one another without any explanation. Those which, in the author's judgement, are the most reliable have been included.

The political context

The history of housing in Caracas cannot be understood outside the political context of the country. With only a brief interruption between 1945 and 1948, Venezuela was ruled continuously by military governments from independence until 1957 (Carrera Damas 1983, Ewell 1984, Martz & Myers 1987). Since 1958, the country has been one of the most democratic in Latin America, at least in terms of the regular interchange of power between civilian regimes. Power has been contested by two major parties, Democratic Action (AD) and the Christian Democrats (COPEI), which have increasingly monopolized the vote cast in the country. In 1958, they gathered 64% of the vote; in 1988, 96%. Both parties have occupied the presidency (Table 7.1), although AD tends to be the majority party.

While the Venezuelan political system is democratic, it is also highly exclusionary (Gil Yepes 1981) and corrupt (Hellinger 1991). Power is highly centralized, both functionally, through the president and his party, and geographically, because decision-making is strongly concentrated in Caracas (Stewart 1987). The state listens to corporate interests which petition the state through the offices of the major parties. It listens in much more limited fashion to popular demands, although political rhetoric is highly populist in tone.

Stability has been maintained by patronage dispensed through the state bureaucracy and the party political machines. Indeed, patronage has been the cement that has held the Venezuelan form of democracy together. Since the 1940s petroleum revenues have provided the government with ample means by which to win friends and gain votes. One consequence has been that large sums have been channelled into social expenditure, including the construction of a considerable amount of social-interest housing. Of course, the sums available have never been sufficient to reduce the housing deficit and, in response, the political parties have sought the support of the poor through less expensive and more covert policies. In particular, they have won their votes by covertly encouraging the invasion of state, and sometimes even private, land. Once established, few invasions have been displaced and most have eventually been serviced from the petroleum revenue inflated social budget. While much of the activity has been conducted with questionable efficiency, it has meant that infrastructure and services have been introduced into most low-income areas of the city (Marcano 1981).

The demographic and physical context

The housing situation in Caracas cannot be understood without reference to the rate of urban growth and the location of the city. Caracas grew extremely rapidly from 1936 to 1971 (Table 7.4). Such growth was encouraged by a highly concentrated pattern of government expenditure and the centralized pattern of political decision-making (Gil Yepes 1981, Stewart 1987). While there has been a strong tendency since 1960 for manufacturing to locate westwards along the Caracas–Puerto Cabello corridor, the capital continues to accommodate a considerable proportion of Venezuela's economic activity. In 1985, 63% of commercial bank deposits were captured in the metropolitan area (COPRE 1987) and, in 1986, 49% of industrial establishments were concentrated in the capital region (OCEI 1987). As a result, the economy of Caracas has continued to expand despite the physical problems of accommodating further growth.

In Caracas, where the city has developed along the valley of the River Guaire and its tributories the Valle and the Baruta (Fig. 7.2) and is hemmed in by the surrounding mountains, urban expansion is highly constrained. Only 11,000 ha of land are available for urban development, although the neighbouring Tuy and Guatenas-Guatire valleys contain a further 35,000 and 5,000 ha of land,

Table 7.4 Population of Caracas, 1936–90.

Year	Population	Annual growth rate
1936	235,160	
1941	324,317	6.6
1950	693,896	8.8
1961	1,336,464	6.1
1971	2,183,935	5.0
1981	2,583,396	1.7
1990	2,989,219	1.5

Note: The figures do not include La Guaira or other coastal parts of the Federal District. Nor do they include parts of the State of Miranda, which grew rapidly between 1971 and 1990, and which are now arguably part of the Greater Caracas area. Such areas include the *municipios* of Guaicapuro, Urdaneta, Lander, C. Rojas, S. Bolívar, Paz Castillo, Independencia and Zamora. If all of these areas are included in the 1990 figure, the total Caracas population is 4,101,494 and the annual growth rate between 1981 and 1990 rises to 2.2%.
Source: Negrón (1982) for 1936 to 1981, and preliminary census figures for 1990 presented in OCEI (1991).

respectively (OMPU 1972b). As a result, the urban area of Caracas has expanded rapidly, and Greater Caracas now arguably occupies an area which stretches from the Caribbean coast through the Guaire Valley to Los Teques in the southwest and the new towns of the Tuy Medio in the southeast. Within the central valley, the combination of the physical shortage of land, a capital-intensive building industry, and the massive injection of petroleum revenues into the built environment has created a densely populated, high-rise pattern of urban development combined with the proliferation of shanty-towns, a feature permitted by political populism (Table 7.5).

Table 7.5 Physical development of Caracas, 1897–1985.

Year	Total hectares	Area of barrios (hectares)	Gross density of city
1897	430[1]	na	180[1]
1936	542[1]	na	430[1]
1949	na	750[1]	na
1950	4,586[1]	na	na
1959	8,000[1]	1,067[1]	151[4]
1966	12,000[2]	2,434[2]	na
1971	15,000[2]	2,973[2]	145[4]
1974	16,250[2]	3,231[2]	na
1979	19,750[2]	4,043[2]	127[4]
1985	na	4,159[3]	na

Sources: [1] OMPU (1972a); [2] Perna (1981); [3] FUNDACOMUN (1985); [4] Author's estimates.

Table 7.6 Types of housing in Caracas, 1950–81.

	1950	1961	1971	1981
UNITS				
Ranchos	24,613	51,010	67,202	68,555
Houses	75,910	83,511	149,309	240,967
Apartments	18,843	83,451	160,431	254,811
Rooms	–	29,372	22,719	–
Condominiums	5,137	4,419	–	–
Total	126,867	256,526	399,661	564,333
PERCENTAGES				
Ranchos	19.4	19.9	16.8	12.1
Houses	59.8	32.6	37.4	42.7
Apartments	14.9	32.5	40.1	45.2
Rooms	–	11.4	5.7	–
Condominiums	4.0	1.7	–	–
Total	98.1	98.1	100.0	100.0

Sources: 1950 and 1961 adapted from Myers (1978), 1971 from the census; and 1981 from OCEI (1986).

Caracas: Growth of Urban Area 1567 -1988

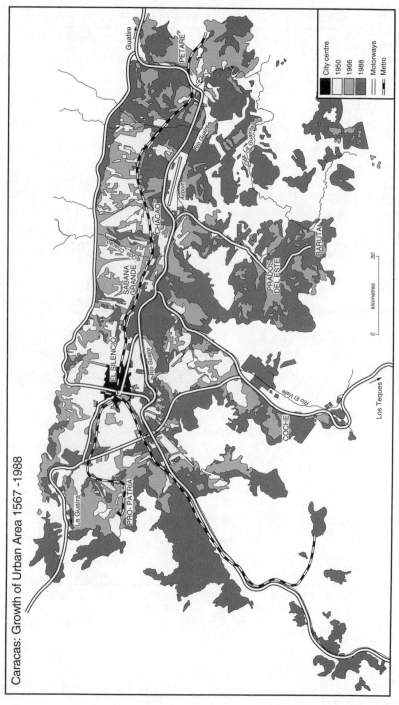

Figure 7.3

Table 7.6 reveals how this has affected the nature of the city's housing. As the city's growth has been increasingly limited by the amount of land available, the proportion of families living in apartments has risen consistently through the years. The city's topography has also affected the shape of city, urban growth spreading lineally along the valleys, linking previously independent towns into the increasing urban sprawl (Fig. 7.3). These subcentres, linked by a motorway system since the 1950s and by a two-line metro system since 1987, make the urban form of Caracas very different from that of either Mexico City or Santiago. Caracas, with its subcentres and its lineal form, has a highly distinctive transportation pattern and particularly severe congestion problems.

Table 7.6 arguably underestimates one further consequence of the land pressure, the growth of the proportion of Caraqueños living in self-help settlements. Table 7.7 provides an alternative picture, showing that the proportion of the population living in *ranchos* has increased remarkably through time. While the definition of *ranchos* is extremely vague, and Table 7.7 certainly exaggerates the proportion of the population living in bad housing conditions, the growth of semi-serviced *ranchos* is an undeniable feature of the growth of the greater Caracas area. Many of the families living in such accommodation are, of course, tenants.

Table 7.7 Growth of *ranchos* in Caracas, 1961–90.

Year	Population living in *ranchos*	Percentage of total Caracas population
1961	280,000	21
1964	556,000	35
1971	867,000	39
1985	na	61
1990	2,685,000	60

Notes: The 1990 figure is based on the urban population in the Capital Region, which includes the whole of the Federal District and the State of Miranda.
Sources: 1961 and 1964, United Nations (1979); 1971, OMPU (1972a); 1985, FUNDACOMUN (1985); 1990, CNV (1991).

State intervention in rental housing

As in most other Latin American cities, rental housing was the traditional form of tenure for most of the urban population until very recently. Rental housing accommodated a majority of the population of the Federal District, even in the 1950s (Table 7.8). Thenceforth, the tenant population declined as more households shifted to owner-occupation, both in the formal and in the informal sectors. The relative proportion of tenants fell in the Federal District from 55% in 1961 to 30% in 1981. While the relative number of tenants fell dramatically, their absolute numbers continued to increase: in the metropolitan area from 155,000 in 1971 to 176,000 in 1981.

State action has contributed strongly to this shift. First, it has reduced the rental housing available in the central city through its efforts at urban renovation. Secondly, it has discouraged the building of housing for rent, through the introduction of rent controls. Thirdly, it has encouraged owner-occupation among the middle class by giving subsidies both to home owners and to the builders of owner-occupied dwellings, and by financing large numbers of public housing units for sale. Finally, it has allowed owner-occupation to grow among poorer households by permitting families to invade land on the periphery of the built-up area.

Urban renewal programmes constituted an early attack on the traditional rental housing stock in the centre of the city. Flush with oil revenues, the Medina government began to rebuild the El Silencio district

Table 7.8 Housing tenure in Caracas, 1950–81 (thousands of homes).

Year	Owners	Tenants	Other	Total
FEDERAL DISTRICT				
1950	58	57	8	122
1961	102	126	1	229
1971	174	132	11	334
1981	277	132	23	434
PERCENTAGES				
1950	47.4	45.9	6.7	100
1961	44.6	54.9	0.4	100
1971	52.1	39.6	3.3	100
1981	63.9	30.4	5.3	100
METROPOLITAN AREA OF CARACAS				
1971	na	155	na	400
1981	357	176	32	564
PERCENTAGES				
1971	na	38.9	n.a.	100
1981	63.2	31.2	5.6	100

Note: Percentages do not add up to 100% because a number of households did not declare the tenure of their home. This was a particularly serious in 1971.
Source: CEU (1990) for the Federal District; unpublished census results for the metropolitan area in 1981; published data on the metropolitan area for 1971.

in 1941 (Fig. 7.4). The existing "slum" housing was demolished and replaced with low blocks accommodating 207 shops and 747 apartments. The scheme, undertaken by the Banco Obrero, was widely regarded as a great success and formed the basis for later programmes based on the concept of the superblock (CEU & OESE 1977, García & López 1989). The superblocks were constructed just to the west of the central area in the 1950s, the 23 de enero and Atlántico Norte complexes accommodating some 105,000 people and blocking any likely expansion of irregular housing in that part of the city (Perna 1981). Other major schemes furthered the process of urban renovation in the central area. In the early 1950s the construction of the twin towers of the Centro Simón Bolívar administrative complex in El Silencio, the Helicoide exhibition centre, and the Avenida Bolívar motorway with its associated bus station helped turned most of the zone into a specialized, non-residential, central area. Revalorization of the old central area was also helped by major road improvements which linked El Silencio to the rest of the city. Critical among these improvements was the building of Avenidas Sucre, San Martín, Urdaneta and Fuerzas Armadas. Much later, the central business district was physically extended by the construction of the vast Parque Central complex. Begun in 1969, but still under construction, this enormous high-rise business, cultural and residential complex provides 12,500 jobs and accommodates 10,000 residents (Hure de Socorro 1978).

The process of urban renewal involved a great deal of housing destruction. Accommodation which might otherwise have been adapted for rental housing was simply replaced by high-rise commercial and residential development. Both public and private enterprise redeveloped large areas of the central area. Consequently, Caracas possesses few equivalents of the central *vecindades* of Mexico City or the *conventillos* of Santiago.

Rent controls

Rental legislation has also acted as a constraint on rental housing construction. The first form of intervention came in 1921 when the state passed a law to regulate living conditions in *casas de vecindad*.[1] There is little information about its overall impact, or even on the degree to which it was implemented. The impression is, however, that it had little real effect on most kinds of rental accommodation. Today, although the legislation is still in force, it has little effect and there is no current interest in updating its provisions.

Much more significant was the legislative decree on housing eviction (*Decreto Legislativo sobre Desalojo de Viviendas*) which was introduced in

---

1 *Reglamento sanitario de casas de vecindad* 31 March 1921 and its reform in 16 August 1923.

Urban Renovation in the Centre of Caracas

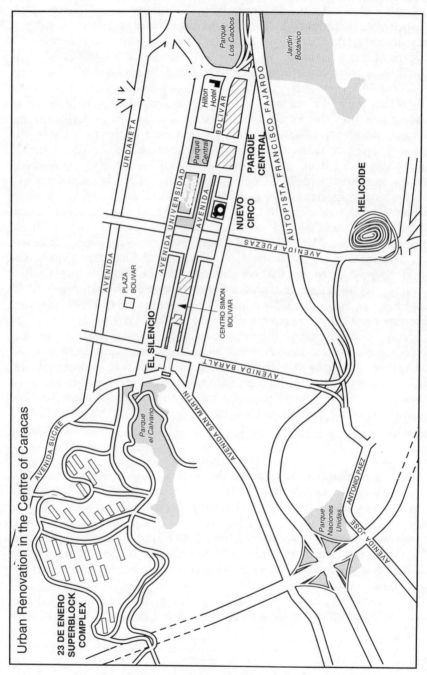

23 DE ENERO
SUPERBLOCK
COMPLEX

AVENIDA SUCRE

Parque
el Calvario

EL SILENCIO

AVENIDA SAN MARTIN

AVENIDA

PLAZA
BOLIVAR

AVENIDA UNIVERSIDAD

AVENIDA

Parque
Central

Parque
Los Caobos

Hilton
Hotel

BOLIVAR

NUEVO
CIRCO

PARQUE
CENTRAL

Jardín
Botánico

CENTRO SIMON
BOLIVAR

AVENIDA BARALT

AVENIDA FUEZAS

AUTOPISTA FRANCISCO FAJARDO

HELICOIDE

Parque
Naciones
Unidas

ANTONIO PAEZ

AVENIDA JOSE

Figure 7.4

1947. The decree was significant because it eliminated the traditional form of private contract between landlord and tenant. Previously the state had hardly intervened in the rental field, and matters such as the level of rent and the question of eviction were resolved individually between landlord and tenant. The new decree laid down the rules under which a landlord could eject a tenant. The main grounds for eviction were: the non-payment of rent; when the landlord required the accommodation for himself or his family; when the building was to be demolished or improved; or in the case of damage to the property by the tenant. The major outcome of this decree was to give the tenant much greater security of tenure. It led, however, to fears among landlords about whether they would be able to evict tenants when they wished.

From the perspective of the landlord, a bad situation was aggravated by approval of the Rent Control Law (*Ley de Regulación de Alquileres*) of 1960.[1] Under the provisions of the new law, the Rental Directorate of the Ministry of Development was responsible for establishing the appropriate level of rent. For newly rented property this was first set as a percentage of the value of the house; 10% up to a house value of 75,000 bolívares (US$17,400), 9% to 150,000, and 8% thereafter. All existing rents lower that 1,500 bolívares would be reduced. The amount varied according to the rent; for rents lower than 600 bolívares the proportion was 25%, for those above 1,200 bolívares it was 19%. The law also made it illegal for property to be let which failed to meet minimum health and sanitary requirements; a provision that was really directed against the renting of shacks (see below). Rodríguez (1976) claims that the 1960 rent controls severely diminished private sector interest in rental housing. Of course, the early 1960s were a period of major recession in the construction industry, and the decline in interest may have been a knock-on effect of the low level of petroleum revenues at the time; nevertheless, it is clear that the rent controls were the subject of vehement criticism from landlords. Perhaps as a result of these complaints, the permitted rent level was raised in 1966 to the equivalent of 12% of the value of the property (CEU 1990).

Despite the modification, landlords continued to criticize the legislation, and real-estate interests still blame the rent control legislation as one of the main reasons underlying their failure to invest in rental property. According to the president of the *Cámara Inmobiliaria* in 1987: "We find that in Venezuela no-one has built rental housing during the past twenty years or so; the result of a lack of incentives for the sector and of the fixing of rent levels."[2] A similar view has been expressed in the press by Ruth Krivoy: "The building of housing for rent is virtually non-existent in Venezuela. Influencing this situation are the laws, the rent controls, the

---

1 *Gaceta oficial* 26319 of 1 August 1960.
2 *Economía Hoy* 24 July 1989.

standards laid down for *inquilinatos*, and the attitude and conception of the business of housing construction which aims at quick sales as an integral element in the profitability."[1]

It should be emphasized that this criticism came from formal-sector landlords. The legislation has never had any real influence in the *barrios*. The only way in which the existing legislation may have influenced conditions in low-income areas is through the prohibition on landlords renting out property that does not meet minimum health and sanitary requirements.[2] In practice, even this constraint has had little effect, because most housing in the *barrios* of Caracas is supplied with water and services and therefore is not automatically excluded from rental arrangements. In any case, the legislative controls over rents would have been ineffective, in so far as most property lacked title deeds and would not, therefore, have a legally defined property value. Thus, in the low-income settlements a market has been operating, unconstrained by the rent control legislation. In the *barrios*, investment in rental property has been occurring despite the rent-control legislation.

While the rent controls may have worried formal-sector landlords and led to some disinvestment, their impact declined rapidly as a result of inflation. In recent years, the rent controls can have had little real effect since accommodation worth more than 225,000 bolívares was excluded from the legislation. In 1960, when the legislation was approved, this figure represented the not unreasonable sum of US$52,300. But as it was never revised, the amount of property excluded by the legislation clearly rose dramatically through time. By 1987, when a new rental law was introduced, all property worth more than US$7,260 was excluded from the controls. At that time, the only new property worth less than that figure could not have been let because it would have constituted a slum!

However, one important qualification should be made; on older property the rent controls could be maintained by the use of the system of the *traspaso*. Cilento (1989) notes that under this mechanism existing tenants are able to pass on to others the difference between the controlled and the regulated rent. As a result, a widespread black market has emerged.[3] The essential problem lies with the power given by legislation in 1972 for tenants to stay beyond the period of their contract without the danger of eviction. The difficulty of evicting a tenant is clearly a great disincentive to the landlord although, given the low maximum value of the property, it can hardly be seen as a major disincentive to the construction of new housing for rent. In any case, there are ways around, even the inconvenience of *traspasos*. Some landlords have modified their

---

1 *Economía Hoy* 24 July 1989.
2 *Gaceta oficial* 26319 of 1 August 1960.
3 Gil (1991).

contracts so that they can remove tenants through the civil courts rather than going through the administrative system set up to arbitrate between the landlords and their tenants.[1]

Despite the vehemence of the complaints from the real-estate lobby, therefore, the rent-control legislation has merely contributed to the decline in rental housing investment rather than representing the major cause.

Until recently, there has been little official regret about the decline in rental housing. However, during the 1980s the economic recession led to a severe fall in the capacity of most Venezuelan families to buy their own home. The combination of falling incomes and high interest rates, together with continuing high land and property prices, meant that few, even among the middle class, could afford to buy a home. While the state continued to seek ways to increase owner-occupation (see below), it also began to try to stimulate the construction of housing for rent.

In 1986, for example, Decree 1280 was introduced in an attempt to reactivate the construction industry and to increase the supply of housing throughout the country. Some 5 billion bolívares from IVSS pension funds were diverted into the SNAP system (see next section). Most of the money was to be offered as 20-year mortgages at a fixed rate of 9% on property up to a value of 3.5 million bolívares (US$175,000). However, builders who were forced to let housing because they could not sell it after 1 year on the market were also offered subsidized mortgages (Cilento 1989).[2]

In addition, the Rent Law was reformed by Decree 1493 in January 1987. The aim behind the modification was to encourage large and small investors to put money into rental housing. Landlords with established tenants could charge rents of between 8% and 10% of the value of single or semi-detatched properties up to a value of 225,000 bolívares; rents in multifamily buildings could be as much as 12% of the value of the property. Perhaps still more significant, however, was that new contracts would be exempt from rent controls. In practice, rent control on new property disappeared in Caracas.

However, the new measures hardly unleashed a great wave of building for rent. Because of the recession and the great slump in the construction industry, there was very little building of any kind (Table 7.3). The cheap mortgages did not stimulate a great deal of interest from the construction and banking sectors because of the high risks involved,[3] and the removal of rent controls merely encouraged speculation, failing to convince new investors to generate the new housing that would reduce the housing

---

1 By introducing the clause *Letras de cambio* into the contract, the landlord avoids all the rental control legislation. Rather than passing through the *Direccion de Inquilinato* and then to judgement within its tribunals, which takes an interminable period, he can proceed directly to the civil courts, *Tribunales civiles-mercantiles*.
2 The mortgages were on the same 20-year 9% fixed-interest basis.
3 *Latin American Regional Reports, Andean Group* 9 October 1986.

shortage and lower rents in the medium term (Cilento 1989).[1]

As a result, a further series of incentives were introduced in September 1989 when the Housing Policy Law was approved.[2] Again conceived principally as a means of reactivating the economy while ameliorating the housing situation, the Law dedicated 5% of the Federal government's ordinary budget to housing, and made it obligatory for 3% of every worker's salary to be deposited in the mortgage bank system. This huge injection of funds was to be devoted to building more houses for lower-income groups. Loans up to 400 minimum salaries were available in Caracas. With respect to rental housing, a new system of subsidized credits was made available to the builders of homes for households earning less than three minimum salaries. The effect of this legislation cannot yet be evaluated.

In 1990, the government also submitted a new Renting Law to Congress which promised to remove all forms of rent control gradually over a period of several years. This proposal has stimulated considerable controversy and so far it has not been approved. Whatever the outcome of the new proposals, the abolition of rent control will have little effect in the poorer areas of the city. As the next chapter shows, renting in the *barrios* is affected little by the intricacies of Venezuelan law, and nothing has been said in the public debate about conditions in the low-income neighbourhoods. Removing rent controls will have no effect because they do not operate now.

Official encouragement for owner-occupation

If the state has been responsible for the decline in rental housing, it has been due primarily to the encouragement that it has given to owner-occupation since 1958. Since that time, the "state has been progressively encouraging, both implicitly and explicitly . . . the need for all Venezuelans to own their own property" (CEU 1990). The offer of credit on generous terms has convinced most households in the formal sector that they should purchase their own home. In turn, this attitude strongly discouraged investors from building accommodation for rent. At the same time, the state has built public housing only for sale and not for rent.

---

1 It is estimated that rents rose between 25% and 94% according to zone between July 1989 and July 1991 (*El Nacional* 23 July 1991).

2 *Ley de Política Habitacional* was approved in September 1989 and began to operate in January 1990.

The private sector

Strong support from the United States government encouraged Venezuela to establish a savings and loans system in the late 1950s (CEU & OESE 1977, Myers 1978). The new Venezuelan system was based on a cross between the British building society and the US Savings and Loans Associations. USAID provided both technical advice and capital to support the establishment of the new system. In 1961, the Betancourt administration created a Commission within the Banco Obrero to control the specialized activities of ordinary commercial banks, mortgage banks and finance societies (Cilento 1989). The system was modified in 1966 when the Banco Nacional de Ahorro y Préstamo was established and the Sistema Nacional de Ahorro y Préstamo (SNAP) created to help generate more housing for lower-income groups.

The new loans system was financed with capital from the oil companies. It provided credit over much longer periods than had previously been available and established subsidized interest rates for "social-interest" housing. The system expanded rapidly, with the number of private mortgage banks increasing immediately from one to four, and to 10 by 1977. The value of loans increased from 59 million bolívares in 1959 to 4.86 billion in 1973, and to 32.9 billion 10 years later (CEU & OESE 1977, BCV 1984). Between 1979 and 1983 slightly less than half of all loans were registered in Caracas (BCV 1984). There is no doubt that owner-occupation became widespread in the capital as a result of this system.

There have been two major problems. The first has been the impact of the great swings in construction activity, linked to changes in the state of the national economy. Second, there has been the problem of keeping the price of housing affordable by the majority of the population.

The fluctuations in construction activity since 1970 have been dramatic. From 1974 to 1978, there was a spectacular boom in production and prices, between 1983 and 1985 a major slump, and after 1986 a further rise before activity plummeted again in 1989 (Table 7.2). The boom of the 1970s was caused by the huge rise in petroleum revenues; the slump of the 1980s by the subsequent fall, the onset of the debt crisis and the fall in real incomes. The recession since 1982 has affected the construction industry very badly and the government has not been very effective in overcoming the crisis, despite many attempts at reviving the sector (Cilento 1989). Indeed, the introduction of fixed interest rates at a time of rapid inflation in 1986 seems only to have fuelled inflation in the housing sector.[1]

---

1 Decree 1280 of 24 September 1986 established a direct credit system with fixed subsidized rate of interest for 20 years, to encourage social-interest housing. The plan was to build 25,000 units in the 160,000 to 250,000 bolívar price range. The Venezuelan Social Security Institute (IVSS) was required to finance the scheme by buying up to Bs.5 billion in mortgage certificates from the mortgage banks and savings and loans associations (Lloyds Bank 1986).

Although other factors have influenced the situation, notably the dollarization of the Venezuelan economy, the subsidized interest rate has stimulated speculation. Unfortunately, it has not led to an increase in the amount of residential construction.

The second problem has been to keep the price of housing affordable. The state has attempted to do this by building public housing (see below) and, increasingly, through subsidizing the purchase and construction of social-interest housing. Indeed, the introduction of the SNAP system in 1966 to complement the activities of the ordinary mortgage banks was motivated principally by this desire. The system has used a variety of methods over the years to keep prices accessible to targeted groups of the population. In 1969, for example, tax relief was offered to companies that built or financed housing for the middle class, and the following year the Banco Obrero offered to buy unsold units worth less than Bs.45,000 (US$10,463). In 1981, a subsidy was introduced (Decree 1134) which was effectively an interest-free loan to the owner on part of the debt of social-interest housing. The attempt to encourage the production of lower-cost housing has been a recurrent theme, although the real price of that housing has persistently risen. The system has been undermined both by the high land values in Caracas and other Venezuelan cities and by the state of the economy. In practice, the SNAP system gradually broke down. By the late 1980s, no social-interest housing was being built in Caracas (Gil 1988) and the system as a whole had become virtually indistinguishable from the private mortgage banks (Cilento 1989). Both systems faced major problems in attracting savings; fixed interest rates during a period of rapid inflation increasingly encouraged the illegal use of subsidized funds for speculation. Most private money was used for speculation in luxury housing; the private sector was catering only for the top 5% of the market (Cilento 1989).

Public housing construction

The Venezuelan government first began to construct public housing in 1928 when the Banco Obrero was established. Between 1928 and and 1958 this agency built most of the country's public housing, a total of 41,000 units (Table 7.9). Very little of this housing was for rent. The bank did build some rental housing in the 1930s, but most was outside Caracas. After 1946, however, the agency sold all of its new housing and disposed of its existing rental housing stock.[1]

Three-quarters of all the public housing built before 1958 was built in Caracas. Most of this housing was in the form of apartments in the super-

---

1 The Banco Obrero encouraged payment of a sum over and above the rent which would go towards the 5% downpayment on the house. At that point the rental contract would turn into a mortgage.

Table 7.9 Public housing construction in Venezuela
and the capital region, 1928–90.

Period	National total	Capital region	%
1928–58	40,675	31,224	76.8
1959–63	33,629	4,855	14.4
1964–68	103,092	19,720	19.1
1969–73	172,102	35,868	20.8
1974–77	119,568	11,755*	8.3*
1980–83	107,792	20,078	18.6
1984–88	168,281	24,845	14.8
1989–90	40,224	4,656	11.6

* Only includes the Federal District.
Source: Banco Obrero (1973) for 1928–73; Handelman
(1979) for 1974–77; CNV (1991) for 1980–90.

block programme of the Pérez Jiménez regime. It is estimated that, in
1978, one in six *caraquenos* lived in INAVI housing (Handelman 1979),
mainly in the notorious 10,000-apartment *23 de enero* development built in
the 1950s by Pérez Jiménez and the 25,000-apartment Caricuao complex
built in the 1960s.[1]

Table 7.9 shows, however, that very little housing has been built in the
capital region since the 1950s. Governments have directed the bulk of
construction towards the provincial cities, mainly because of the high cost
of building in the Caracas region. This tendency has been accentuated,
however, by the strong move away from large-scale building schemes
towards more "progressive" programmes such as sites and services. For
this latter kind of programme the extremely high cost of land in Caracas
has been prohibitive.

The move away from large-scale building schemes was first attempted
under the first Democratic Action government. FUNDACOMUN was
established in 1962 both to improve the quality of local government and
to help channel funds from USAID into a viable slum-improvement
programme. In Caracas, a Committee for Remodelling the Barrios was
established; an approach intended both as an antidote to the superblock
and as a means of winning influence among poorer voters (Myers 1978).
Neither objective was very successful, and by the time the next
administration took over "self-help housing programs for the Caracas poor
were seen as politically counterproductive, technically deficient, and
corrupt" (Myers 1978). A return to conventional public housing policies

---

1 The Banco Obrero was restructured in 1974 and renamed the National Housing
Institute (INAVI).

was the result, most spectacularly shown in the Caricuao complex in the southwest of the city. Further efforts at improving self-help housing were made by the Caldera administration and by the subsequent Pérez government with its *barrio* improvement programme. None of these efforts, however, had much effect in Caracas in terms of creating new popular housing areas. They were reactive policies intended to improve conditions in existing settlements (see next section).

During the 1980s, relatively little has been done to increase the supply of housing through schemes such as sites and services. In addition, little public housing has been built in Caracas. CEU & OESE (1977) were complaining at the end of the 1970s that one-quarter of the poor of Caracas could not afford the cheapest form of public housing, and a decade later Lovera (1987) was lamenting that even "so-called 'social-interest' housing intended for the middle class shone through its absence".

In practice, therefore, Venezuela has failed to resolve its housing problems through conventional building and finance methods, despite a high level of per capita income by Latin American standards. As Cilento (1989) expresses it: "In spite of the effort made by INAVI and the SNAP . . . housing conditions for those with low and medium incomes have declined substantially. And this has happened despite a decline in the rate of population growth and in internal migration, despite the hundreds of thousands of millions of bolívares devoted to residential investment, and in spite of the fiscal sacrifice caused by tax relief and incentives and in spite of the self-help contribution of the *rancho* dwellers."

Self-help housing

The solution of the low-income housing problem has been left to the poor themselves. They have built their own self-help homes on land that they have invaded. Most of this land has been located on the hillsides of Caracas; land that was too expensive for formal-sector construction. The illegal occupation of land has been occurring for years. Several invasions certainly occurred during the 19th century, and Stann (1975) estimated that *ranchos* may have accounted for 10% of the total buildings in the city in 1891. But, the real beginning of the self-help process was in the 1920s (Ramírez et al. 1991), when the expansion of regular bus services begain to open up new areas for settlement (Stann 1975), and the period of greatest expansion came with the brief interlude of democratic rule in 1945. It is estimated that 63 *barrios* were established in the 1940s, 36 of them in 1947. The process was slowed by the return of military rule and the removal programmes of Pérez Jiménez. Nevertheless, the 1950s still saw the formation of 100 more settlements, admittedly most being formed during the Emergency Programme of 1958 and in the first year of the

Betancourt government. By the 1960s, the pace of growth was slowing, although 60 new settlements were still established during the decade, and by the 1970s and 1980s the shortage of free land meant that new settlement formation was declining fast. In 1989 it was estimated that there were 379 *barrios* in the city and 608 in the whole Capital region.[1]

As a consequence of the invasion process, the proportion of Caracas' population living in self-help housing has risen rapidly through time. It was estimated that 21% of the population lived in some kind of informal settlement in 1961, some 39% by 1971, and some 60% by 1990 (Table 7.7). Although the rate of settlement formation has slowed in recent years, between 1978 and 1985 63% of all houses in the metropolitan area were still built without planning permission (Bolívar 1989).

Most of the land occupied by invasion has belonged to the state. A study in 1984 established that 49% of invaded property belonged to the government, 15% to private owners, 19% belonged to a mix of owners, while the owners of 18% of the land were unknown.[2] While some invasions have been organized expressly against the wishes of the owners, most occupations have received the tacit support of some political actor (Ray 1969, Bolívar 1989). The invasion of land has formed a key part of the process of winning friends and influencing people, and all of the main political parties have used land as a form of gaining political support (Bolívar 1977, Gilbert & Healey 1985). Indeed, land seems to have been used to reward both the invaders and the invaded. As Pérez Perdomo & Nikken (1982) point out, "due to the State's generalized lack of action, the landowner loses his ownership rights but, due to its selective action, favoured landowners regain their rights through expropriation in the public interest.". While a few politically weak owners do lose out, most manage to negotiate some kind of compromise (Gilbert & Healey 1985).

In Caracas, few invasion settlements have been displaced since the 1950s. Since the re-establishment of democracy, displacement has occurred only when particularly sensitive areas have been occupied, where the political backer proves to lack sufficient friends in high places, or where the settlement has been founded in a particularly dangerous location. As a top municipal official in Sucre puts it, "every uncontrolled occupation becomes a tolerated occupation, save those invasions of municipal land which are manifestly unstable".[3] Indeed, officially most removal programmes have always been justified in terms of the geological

---

1 The estimate was made by Dra. Xiomara Alemán, Coordinadora del Sistema de Información y Análisis de barrios, FUNDACOMUN and is cited in CEU (1989). The figure for the Capital region is based on FUNDACOMUN estimates (CNV 1991).
2 The study was conducted by the Gobernación del Distrito Federal, ORCOPLAN and FUNDACOMUN.
3 *Contraloría Municipal del Distrito Sucre – Estado Miranda (1985) Estudio Jurídico sobre uso y administración de la tierra pública municipal del Distrito Sucre del Estado Miranda, Petare.*

conditions, frequently citing the danger of landslides (PREALC 1981, FUNDACOMUN 1988).

Despite the fact that most poor households live in settlements founded by invasion, a minority of these households are themselves invaders. Many families have bought into the settlement after formation; indeed, in some settlements this is a majority of the population. In addition, there are many tenant and sharer families. The irony is that although the law forbids the building, renting or selling of *ranchos*, little is done to apply the law.

As a result, there is little fear among the *rancho*-dwellers that they will be evicted. Security of tenure is not a problem, even though few hold title deeds (Karst et al. 1973, Pérez Perdomo & Nikken 1982). Given relatively high incomes, therefore, most self-help housing in Caracas is well built and serviced. As Ramírez et al. (1991) note, "The dominant landscape is one of rather large, solid houses built with bricks and other industrially produced materials, concrete surfaced roads and basic facilities. Although they have been built illegally, on invaded land, very little is temporary or precarious in these squatter areas." The *ranchos* have adapted their form to the terrain, producing a labyrinth of streets and alleys on the hillsides, somewhat reminiscent of a mediaeval landscape (Negrón 1987).

Settlement consolidation has been helped by the fact that many relatively affluent families live in self-help housing in the city; families excluded from the formal sector by the high cost even of apartments. But if, until 1989, the *barrio* population was hardly poor by Latin American standards, many have long been forced to live in difficult environmental conditions. Many of the hillsides on which the *ranchos* are constructed are unstable, and there are frequent landslides during the wet season. Certain parts of the city also suffer from severe air pollution, for example, smoke and dust from the La Vega cement plant affects an area containing at least 300,000 people.

Clearly, however, the state has been heavily involved in the improvement of the settlements, and particularly in their servicing. The first concerted effort to service the *barrios* came with the establishment of the *Programa de Remodelación de Barrios del* AMC in 1960. This plan aimed to remove the worst settlement, service the better-located *barrios*, and house the less poor households in social-interest housing (Marcano 1987). Unfortunately, as we observed above, it was not a great success. Of greater importance was the foundation of FUNDACOMUN in 1958, with its injection of funds from USAID and its efforts to upgrade low-income settlement.

Whether a systematic campaign to improve conditions in the settlements has been underway or not, the Venezuelan state has consistently provided something in the way of services and infrastructure to the majority of low-income settlements. Some neighbourhoods might be excluded for political reasons, or because they occupy private land belonging to significant political actors, but the majority of settlements eventually receive state assistance.[1] This is made clear by the results of the inventory carried out by FUNDACOMUN in 1974-75. According to that survey, 75% of the *barrios* in Caracas had access to piped water, 69% were linked to the sewerage system, 81% had public lighting, and 97% were supplied with domestic electricity (FUNDACOMUN 1978). Admittedly, many of the services were not wholly reliable, the supply of water often being subject to interruptions. Certainly, too, the quality of infrastructure provision is much lower than that supplied to the normal urban development in higher-income areas (Bolívar 1977). Nevertheless, the provision of services and infrastructure to the *barrios* has been regarded as normal procedure by the service agencies ever since 1958. Moreover, services have usually been provided without charge, either for installation or for the subsequent service (PREALC 1981). As Marcano (1981) pointed out, the whole process of demand-making distorts rational allocation methods and leads to inefficiency throughout the metropolitan area. Nevertheless, "the regular system of piecemeal responses to mediated but organized requests has . . . been successful in maintaining a level of servicing of *barrios* which is high by comparison with other South American countries" (PREALC 1981).

Of course, this situation is hardly appreciated in the *barrios*, where the population is only too well aware of the unreliability in service delivery, the regular shortages of water, the power cuts during storms. As such, there is always pressure on the political parties to improve conditions. Frequently, new administrations launch special programmes to respond to these demands. The Caldera administration (1969-73), for example, made a significant effort to provide credit to the low-income settlement, to introduce infrastructure and services and to provide technical help to housing perched on the hillsides (*viviendas en pendiente*) (Quintana 1987). Similarly, while housing was not a high priority for the succeeding Pérez administration, it still introduced a major programme to improve service delivery. Decree 332 of 1974 ordered an inventory of the *barrio* population and introduced a plan to provide services to all "stable" settlements.[2] The service programme was given large sums of money and led to the construction of service modules in many *barrios* in Caracas, and significant improvements in a much more restricted number of settlements (PREALC 1981). Ultimately, however, it failed to transform the servicing situation; indeed, much of the programme's expenditure was already included in existing agency budgets (PREALC 1981).

During the 1980s, new initiatives were announced to rectify the *barrio*

---

1 See the discussion in Gilbert & Healey (1985) for the case of Valencia. See also Bolívar (1989).

2 *Programa para el Ordenamiento de las áreas ocupadas por los barrios pobres de las ciudades del país.* The programme was given considerable clout by the fact that the new president of FUNDACOMUN was the daughter of Carlos Andrés Pérez.

situation. The Lusinchi government ordered FUNDACOMUN to conduct a new inventory of the *barrios*, and in 1988 the Organic Law of Urban Ordering recognized the *barrios* as an undeniable feature of Venezuelan life.[1] For the first time, urban plans were required to incorporate self-help settlements into the urban structure. INAVI announced a new programme to consolidate the *barrios* and, at the local level, the government of Caracas began a campaign to legalize and regularize the *barrios*. In 1988, it announced a bill to give title deeds to the inhabitants of the *ranchos*.[2]

In practice, of course, too little help has been provided in terms of resources. The familiar process of improving the look of the *barrios*, through the building of staircases and façades, while failing to resolve more fundamental servicing and infrastructure problems, has been re-enacted. Given the current economic crisis, it is to be expected that the government will continue to respond on an *ad hoc* basis, making decisions on the basis of a mixture of technical and partisan political criteria.

No doubt, too, governments will continue to face two specific problems. The first is the recurrent difficulty of dealing with settlements that occupy dangerous areas of land or areas intended for major public works. This problem, which was addressed in Decree 332 and in most other government declarations in recent years, re-emerged in the discussions about the recent Organic Law of Urban Ordering. Articles 50 and 51 of this law specifically state that certain *barrios* would inevitably have to be removed because they are in difficult geological areas, or because they interfere with public services and infrastructure. Most significantly, they decreed that such settlements would receive nothing in the way of indemnification payments, removing the main method by which Venezuelan governments had previously managed to relocate settlements with a minimum of fuss.

The second problem relates to the question of title deeds in low-income settlements. This has always been a difficult issue, especially where private land has been occupied or where there are doubts about ownership (Gilbert & Healey 1985). This is demonstrated by the responses of local government in different metropolitan zones. In the Vargas District of the Federal District, where ownership of invaded land is generally clear, the municipal government has been able to sell or rent the land to the settlers. In the Libertador District, the title situation is more complicated, and in settlements in the La Vega area there are many conflicts over title dating back to the time of the *resguardos*. Since 1977 the Council has banned sales until it has established the limits of its patrimony.[3]

---

1 By January 1990 this inventory had still not been completed in Caracas (CEU 1990).
2 INAVI (no date) *Programa Consolidación de Barrios*, World Habitat Day – Workshop on Planning of Technical Assistance.
3 These complications have even led to paralysis in granting rental contracts; a process which was slow even when it was permitted, taking 3–4 years to pass through the municipal system.

Meanwhile, in the State of Miranda, the Council of Sucre has clear ownership to very little land. Most of the invasion settlements occupy *ejido* land, areas over which third parties often claim rights, or which are legally reserved for some specific function. The overall result has been that little land has been sold; only where some overriding political factor has operated has land been sold. As usual with government decision-making in Venezuela, there is little in the way of a clear set of rules.

No doubt government policy in the future will continue to be resolved mainly on the basis of compromise, in the way that it has been over the past 30 years. The fundamental element in this compromise is that the state accepts the reality that poor Venezuelan families accommodate themselves through the self-help process. The state tries to improve and service these families, and introduces regular initiatives to speed up the process, while always retaining the right to remove them (Bolívar 1989).

CHAPTER EIGHT

Review of the
Caracas survey results

Settlement selection

The questionnaire survey was conducted in six settlements in Caracas (Fig. 7.1). Two new self-help settlements were included: Las Torres, a neighbourhood located in the southwest of the city, and Los Chorritos in the east. Two older and highly consolidated settlements were selected: El Carmen de la Vega in the southwest of Caracas and El Carmen de Petare in the east. Two areas of rental housing were also selected in the central district, located in the parishes of San José and San Juan.

Three of the self-help settlements were formed through a process of land invasion, the dominant form of *barrio* foundation in Caracas. In the fourth settlement, plots were acquired through illegal subdivision. Two of the self-help settlements were located in the southwest of the city and the other pair in the east. An important factor underlying this selection was that one new self-help settlement and one consolidated settlement should be located within the Federal District, the other pair of self-help settlements within the administrative jurisdiction of the State of Miranda.

Las Torres is located on a mountainside between La Vega and Los Jardines del Valle. It was formed through invasion in 1983, with new families moving into the settlement gradually over the years. The 100 or so houses are situated either side of the earth road that leads to a power-plant. While some of the houses are built out of concrete blocks, the majority are built out of wood or waste materials, and most have a zinc roof. At the time of the survey the settlement had virtually nothing in the way of communal services, although some infrastructure had been installed. While most homes still lacked water, 80% of the houses had electricity.

El Chorrito is located near to the centre of Petare. A few people had invaded plots in the early 1970s but the settlement did not begin to attract families in large numbers until the 1980s. Most of the sample population moved in during 1987 and 1988, either through invasion or through buying plots from existing settlers. The settlement is located at the bottom of a ravine near to La Dolorita. At the time of the survey, the settlement had 102 houses, most of which were built of flimsy materials; 90% had zinc roofs and one in six still had earth floors. Basic services were available, although one-tenth of the households had an illegal electricity supply and one in six houses lacked a piped water connection even to the plot. The settlement's only advantage was that there was a frequent public transport service along the Petare–Santa Lucía road.

Interviews were also conducted in two consolidated settlements. The first, known as El Carmen, is located in the southwest of the city in the parish of La Vega. It forms part of the administrative area of the District of Libertador, within the Federal District. A few families have been living in the area since 1932 but the real occupation of the settlement began in 1958. The land was originally invaded, although three out of five households claimed to have bought plots from third parties. At the time of the survey, the settlement occupied an area of 29ha and contained at least 1,500 houses. The settlement is located on irregular topography, with many of the streets climbing the mountainside. Many of the homes can only be reached by stairways and are inaccessible by vehicle. Much of the lower part of the settlement has paved roads, all homes have metered electricity and the vast majority have both piped water and sewerage.

The second consolidated settlement is also called El Carmen and will be called El Carmen de Petare in this chapter. Although a few inhabitants were living here as early as 1940, the settlement really dates from 1948. It was founded through illegal sale of land belonging to a mixture of owners (FUNDACOMUN 1978). At the time of the survey most of the population had purchased their plots from third parties, with the rest having received the plot as a gift or through inheritance. It is situated on irregular terrain, and while the wide main streets are paved, many homes are located along winding alleyways which can only be reached by foot. The quality of construction varies, although nearly every house is built of concrete blocks, many still have zinc roofs while others have three or four storeys. Infrastructure is good and nearly every house has metered electricity, water and sewerage. By 1988, the settlement was densely populated; occupying some 10ha of land, it contained at least 1,200 households.

Interviews with tenants were also conducted in a number of *casas de vecindad* in two areas of the central city. Located in the parishes of San José and San Juan, both are areas with many high-rise buildings. The *vecindades* are mainly located between skyscrapers, on plots too small for high-rise development. Lacking any real alternative, the owners have decided to subdivide their properties. Most of the buildings are old and many show signs of deterioration. Some, however, have been built more recently, most of flimsy materials with zinc roofs. The main advantage to be gained from living in the two settlements is that they are well serviced and transport time to the central business area is minimal.

The nature of owners and tenants

The survey revealed that owners and tenants in Caracas are different in a number of respects. Owner households contain 1.5 persons more than tenant households, owners having one child more on average than tenants. The additional child is a function of the age of the owners, who are around 5 years older than the tenants, but the larger household size is also due to the higher incidence of extended families among the owners. More surprising is the fact that tenants have higher household incomes than owners. Tenants are more than three times as likely to have been born abroad, an important consideration in a city where as many as one in five of the interviewed households contained immigrants.

These differences between owners and tenants are considerably blurred, however, when the nature of households in different settlements is considered. There is considerable variation within each tenure group. For example, owners in the consolidated settlements are much older than those in the new settlements. Similar differences are apparent in the family structure; there are many more extended families among owners in the consolidated settlements than among owners in the new settlements or among any group of tenants. The characteristics of owners, therefore, differ considerably between settlements.

There are also clear differences between the tenants in different settlements. Tenants in the central city are younger than average, have much smaller households and few have extended families. As a result, there are few consistent differences between owners and tenants. In terms of age, for example, owners in the new settlements are younger than tenants in the consolidated settlements and about the same age as tenants living in the central city.

Income and tenure

Incomes, whether measured in terms of male earnings or total household income, differ between tenure groups (Table 8.1). Surprisingly, tenants earn considerably more than owners. This difference becomes still more marked when per capita income is considered, tenant households receiving 60% more than those of owners. The explanation of this difference in incomes is fourfold. First, the owners are generally older than the tenants, and since many more of the former are retired, they receive lower incomes. Secondly, the incomes of owner families are underestimated by our recording only the contributions of adult children to the household budget, rather than their total income.[1] Thirdly, a higher proportion of female adults work in the tenant households, raising

---

1 Many adult children contribute very little to the family budget, so that the inclusion of their whole income would increase household income excessively.

household incomes. Finally, per capita incomes among tenants are raised by their smaller family size.

Table 8.1 Household income by tenure and settlement, Caracas.

Settlement	Household income	Per capita income	Male income
Owners	4,888	897	4,228
El Chorrito	3,614	674	3,330
Las Torres	5,007	961	4,461
Carmen de La Vega	5,135	882	4,148
Carmen de Petare	6,100	1,138	5,255
Tenants	5,687	1,418	4,912
Carmen de la Vega	5,947	1,304	4,967
Carmen de Petare	5,863	1,269	4,957
San José	5,187	1,900	4,690
San Juan	5,423	1,629	4,903
TOTAL	5,313	1,090	4,515

Source: Caracas survey.
Notes: Male income from main job and additional earnings. Total income includes the earnings of head of household and partner, plus the contributions of other earners to the household budget. Since grown-up children tend to earn much more than they contribute to the household, this figure tends to underestimate the total incomes of owner households because they have more grown-up children. Per capita income is calculated by dividing the previous column total by the average household size.

There are, however, important variations in incomes by settlement, especially among the owners. Male incomes vary considerably, with owners in El Chorrito earning much less than those in the other settlements. Among the tenants, male incomes are very similar and are higher than those among owners, except in El Carmen de Petare. When household incomes are considered, families in El Chorrito are seen to be particularly poor whereas their neighbours in El Carmen de Petare earn more than any other tenant group. By contrast, in the two southwestern settlements household incomes seem comparable. Among the tenants there seems to be much less variation in household income, the only significant difference being that tenant households in the consolidated settlements earn more than those in the central city because they have more earners. When per capita incomes are considered, however, tenants in the central area are far and away the most affluent group.

By contrast with the income data, there seem to be few significant differences between owners and tenant households in terms of wealth indicators (Table 8.2). Tenants are more likely to own a colour television, but owners are more likely to have a car. New owners seem to have very few consumer durables, and central tenants not many more. This overall pattern of similarity, however, is undermined by the differences between settlements. There is a marked difference, for example, between owners in the new settlements and tenants in the consolidated settlements; in El Chorrito, 18% of owners have a colour television, in the consolidated settlements 70%. In part, this is a function of the difference in income, but it is also a sensible response to the problematic electricity supply in the new settlement. Ownership of consumer durables does not differ much between owners and tenants within the consolidated settlements.

Table 8.2 Wealth indicators by tenure, Caracas (% with article).

Article	Owners	Tenants	Total
Radio	77.7	82.2	80.7
Colour Television	55.2	62.7	53.5
Sewing machine	28.5	27.2	30.0
Refrigerator	73.3	79.3	77.3
Car	21.2	16.3	19.6
Bicycle	11.5	9.1	10.9

Source: Caracas survey.

Employment and tenure

Table 8.3 shows several differences between owner and tenant households in terms of employment. Male owners are more likely to work in the construction industry and are much less likely to be engaged in remunerative work; women tenants are much more likely to be engaged in paid work. However, these differences virtually disappear when the characteristics of owners and tenants are considered by settlement.

The proportion of self-employed male workers, for example, varies from 18% among owners in one new settlement (El Chorrito) to 46% in the other (Las Torres); the proportion working in industry is very high in the first and very low in the second. Similarly, there are variations in the work of tenants between settlements. As a result, employment bears no clear relationship with tenure. There is a higher incidence of construction workers (18%) among new owners than among the other groups, but owners in the consolidated settlements had levels no higher than the tenants. The nature of work is linked less to tenure than to location and age of settlement.

Table 8.3 Employment by tenure, Caracas.

Variable	Owners	Tenants	Total
Male employment in industry	25.7	19.8	22.6
Male employment in construction	13.6	7.7	11.4
Males in paid work (%)	93.0	96.5	92.0
Females in paid work (%)	23.1	43.9	31.7
Self-employed males (%)	32.9	36.5	34.6

Source: Caracas survey.

Age and family structure

Table 8.4 shows that owners are generally older than tenants, but there are considerable differences between the average age of owners in the different settlements. Not surprisingly, owners in long-established settlements are much older than those in new settlements; the average age of owners in El Carmen de la Vega was 51 years, compared to only 34 years in Las Torres. The variation in tenant ages was less marked, ranging from 35 years in the central area to 43 years in El Carmen de Petare.

Table 8.4 Age and family structure by tenure, Caracas.

Variable	Owners	Tenants	Total
Number of persons	5.5	4.0	4.9
Number of children	2.7	1.7	2.3
Size of nuclear family	4.2	3.5	3.9
Age household male	38.6	35.4	38.2
Age household female	37.7	34.3	37.2
Household without male (%)	21.7	16.7	19.9
Extended families (%)	38.7	23.9	33.9

Source: Caracas survey.

Owner households are generally larger than those of tenants. On average they contain one child more and there are also more extended families among the owners, although the difference is not large. The size of household varies little among the owners but there are major differences among the tenants. In the central city, the average tenant household contains only 3.0 persons compared to 4.6 in the consolidated settlements.

Family characteristics are linked to age of heads of household, so that smaller, single-headed households are much more common in the central city and larger, extended families more common among owners in the

129

consolidated settlements. The structure of tenant households in the consolidated settlements is very similar to that of new owner households.

Housing conditions and tenure

Housing conditions vary considerably with tenure (Table 8.5). In general, owners live in larger dwellings than tenants, with one-third of the latter living in one room. In compensation, tenants live in better-quality accommodation, with much higher levels of services, a reflection of the tendency for tenants to rent rooms either in the central area or in highly consolidated self-help settlements.

Table 8.5 Housing conditions by tenure, Caracas.

Variable	Owners	Tenants	Total
Households with one room (%)	9.5	39.1	20.6
Households with 2 or more bedrooms	68.8	51.4	63.5
Households with water connection inside house	69.7	92.7	80.4
Irregular electricity supply (%)	48.4	2.2	27.5
No sewage pipe connecting house or plot (%)	11.3	0.0	6.5
Earth or cement flooring (%)	82.7	58.0	71.5
Zinc roofs (%)	65.7	30.1	49.7

Source: Caracas survey.

Too much should not be made of the tenure difference, however, since variations in living conditions between owners in the new and the consolidated settlements are very marked. Whereas the electricity supply is almost fully metered in the consolidated settlements, only 7% have meters in Las Torres and El Chorrito. All but 2% have water piped to the house in the consolidated settlements, compared to 44% in the new *barrios* where 21% have no water at all. Not surprisingly, the difference is most marked in terms of sewerage, with 95% of homes in consolidated settlements having a connection compared to only 37% in the new settlements. Similar variations are shown in the house structure, in the new settlements 13% have earth floors and 92% have zinc roofs, compared to 0% and 34% in the consolidated *barrios*.

In sum, owners in consolidated settlements tend to live in better conditions than most tenants, but most tenants live in superior accommodation than do owners in new settlements.

Preference for ownership

A majority of owners had previously rented or shared accommodation, although almost one in four had previously owned a home. Among the tenants, few had owned either of their previous homes in Caracas. None of the 77 tenants in the central area with a previous independent home in the city had owned it, and in the consolidated settlements only 9 out of 126 had previously been owners.

Table 8.6 Tenure preference among tenants, Caracas.

Preference	%
To share	0.0
To rent	5.4
To own	94.2
Have looked for own home:	
No	55.6
Yes, successfully	3.3
Yes, unsuccessfully	41.1

Source: Caracas survey.

Both owners and tenants expressed a clear preference for ownership. Almost two-thirds of owners giving a specific reason said that the main advantage was having something to pass onto the children. One-quarter more said that they were glad not to have to pay rent; however, 60% did not have any single reason for owning. Among the tenants the preference for ownership was strong, with 94% saying that they would prefer to be owners. Only in centrally located San José was there a significant minority (16%) who preferred to rent, a function of the high number of immigrants in the sample. Table 8.6 shows that almost half of the tenants had looked for their own house or plot, although only a few had been successful.

The preference for ownership is no doubt influenced by the considerable burden that rents place on the household budget (Table 8.7). The mean monthly rent paid by tenant households was 1,198 bolívares (US$34), at a time when the minimum income was 2,613 bolívares. Over the tenant sample as a whole, rent made up 21% of the total income. When this is analyzed by level of income, rent clearly poses a major burden for those earning low incomes. Those earning the minimum wage or less had to pay half of their total household income in rent.[1] Even for

---

1 There is likely to be some underestimating of income among this group, and no doubt some cases where more than two-thirds of the budget is spent on rent is explained by under-recording of contributions by other members of the household.

131

those earning between one and two minimum salaries, the burden of paying 30% of income in rent is considerable. Only for those earning more than two minimum salaries does the burden become reasonable.

Table 8.7 Rent:income ratios, Caracas.

Earnings in minimum salaries	Number of households by percentage of income paid in rent						Percentage by income band
	0-9%	10-19%	20-29%	30-39%	>40%	Total	
<1.0	0	1	3	4	17	25	51
1.0-1.9	2	23	40	17	22	104	30
2.0-2.9	12	21	15	4	1	53	18
3.0-3.9	10	22	7	1	0	40	15
4.0-4.9	5	7	1	0	0	13	12
5.0+	4	3	1	0	0	8	10
TOTAL	33	77	67	26	40	243	25

Source: Caracas survey.

Caracas has a large migrant population. Among the total sample, less than one-third of men and women had been born in the metropolitan area and almost one in five of the sample were immigrants. Migration seems to have a considerable effect on tenure, for immigrants are much more likely to be tenants, 32% of whom had been born abroad. The foreign-born tend to concentrate in rental accommodation in the city centre, where 52% had been born abroad, although they are certainly not uncommon in the consolidated settlements. In the latter, around one-quarter of the tenants were foreign born and in El Carmen de Petare one-third of all female tenants were immigrants.

Rather few immigrants seem to become home owners. Less than 10% of all owners were born abroad and most of these foreign owners are concentrated in a single settlement; in Las Torres, 19% of the men and 25% of the women were immigrants.

Among the Venezuelans, there seems to be little link between migration and tenure. Owners are no more likely to be natives of Caracas than tenants. In sum, immigration seems to be a much more significant factor in tenure choice than internal migration.

Tenure of the parents seems to make little difference to the current tenure of the household. This is no doubt influenced by the fact that the vast bulk of households had parents who owned their dwellings; an outcome of the majority of both owners and tenants having originated from outside Caracas where it is generally much easier to own. Owners were more likely to have had parents who had owned homes in Caracas, but when the extensive presence of immigrants among the tenants is

132

recalled, there is practically no difference between the tenure background of owners and tenants.

Residential movement

Table 8.8 shows that residential movement in Caracas is quite limited. The average current tenure of all households was 10 years, with 31% having lived longer than 10 years in their present home. While tenants had shorter periods in the current home, the average tenure was still 6 years. The tenants with the longest tenure were not those in the central city but those living in El Carmen de Petare.

When households changed accommodation they did not move very far; very few households had ever moved from one end of the city to the

Table 8.8 Years resident in present house, Caracas

Settlement	Mean tenure (years)	Percentage living more than 10 years in home
Owners		
TOTAL	11	36.1
El Chorrito	4	7.0
Las Torres	3	0.0
Carmen de la Vega	18	79.8
Carmen de Petare	17	46.2
Tenants		
TOTAL	6	13.6
Carmen de la Vega	6	12.5
Carmen de Petare	8	19.0
San José	5	14.5
San Juan	5	11.1
TOTAL	10	31.0

Source: Caracas survey.

other. Most owners had moved from previous homes nearby, and most tenants seem to have moved from a previous home in the same *barrio*. Indeed, almost three-quarters of the total households with a previous independent home within the city had moved less than 5km to their present home. Only 8% had moved more than 10km. Table 8.9 shows that tenants moved even shorter distances than did owners: 82% moving less than 5km, compared to 66% among owners.

Very few households had moved to any of the peripheral settlements from the centre of the city. Even fewer had moved from the western suburbs to the Petare area in the east, and none had moved from Petare to the west.

Landlords

Landlords are older than other owners and much older than most tenants. Table 8.10 compares their characteristics to those of other owners. In general, their households are of similar size, but there is a much higher incidence of extended families. They are much older than other owners and are therefore more likely to be living without a partner. Half of the landlords do not have a regular job, a much lower proportion than that of other owners. Among those who continue to work, male landlords are much more likely to be engaged in the construction industry (25%) than other owners. The landlords are considerably more affluent than other owners. This does not derive from the male's income, which is no greater than that of other owners. It comes from having more workers in the family and, of course, from rents.

Most landlords in the consolidated settlements operate on a small-scale, two-thirds of landlords only having one tenant. The largest two landlords had five tenants each, and only two landlords had two properties. The small-scale nature of the business is underlined by the fact that only 8% of tenants paid the rent to an employee of the landlord. Apart from a group of Portuguese merchants, most landlords in Caracas (60%) live in the same settlement as their property, almost half in the same building. Most of the landlords have lived a long time in their current house, on average for 25 years. They have also been renting for some time, with one having been in operation for more than 20 years. One-quarter, however, have been operating for less than 4 years and fewer than one in six for

Table 8.9 Distance moved from last home to present home, Caracas (%).

Settlement	Less than 5km	5-10km	10km or more	Sample number
Owners	66.4	22.9	10.8	223
Tenants	82.3	11.4	6.3	175
Total	73.3	17.8	8.8	–
Number of respondents	292	71	35	398

Source: Caracas survey.

Table 8.10 Socio-economic characteristics of landlords and owners, Caracas.

Variable	Landlord	Owners
Number of persons	5.3	5.5

Number of children	1.85	2.70
Age household male	56.0	38.6
Age household female	54.0	37.7
Extended families (%)	55.0	38.7
Household income	6,703	4,888
Household per capita income	1,255	889
Male income	4,248	4,228
Colour television	73.0	55.2
Car ownership	27.0	21.2
Years in current house	25	11

Source: Caracas survey.

more than 10 years. Most consider the business to be good or acceptable, with only 12% declaring letting to be a bad business. Despite this general approval, none were expanding their property to raise the number of tenants.

Different strategies are clearly employed by the different landlords interviewed. The most common pattern is that followed by families living in self-help homes, where the size of household is in decline. Heads of household become landlords when their children have left home. They normally continue to live in the property, although a few do move to another house nearby. This strategy of "domestic" renting is often a temporary arrangement, if older children want to return then the family ceases to rent the accommodation. The whole rationale underlying the process of letting is far from that of running a business. There is no calculation of profitability, no thought about the return on capital. Renting is merely a means of increasing the income of the household.

A more business-like strategy is followed by those landlords engaged in "progressive commercial renting". This kind of landlord is usually involved in some other kind of business activity, often diverting profits from trade and commerce into rental housing. Such investment seems only to occur where the owner holds the title deed to the land. Without a property title, investing heavily in improving the accommodation is deemed to be too risky. If the landlord were unlucky, the tenants might claim ownership of the property.

A third kind of strategy seems to be employed by landlords with property in the central city. Many are unable to sell the property or to redevelop it and have little real alternative to renting. Rather than building accommodation for rent, therefore, existing property is subdivided; often it is the landlord's previous home. The property is often managed by professional agents and, unlike the situation in the rest of the city, is often let furnished; in San José 18% of tenants rent furnished accommodation.

135

Landlord–tenant relations

There is a lax contractual relationship between most landlords and their tenants. In the survey, only two out of five tenants had a contract and only one in four a registered contract. Of those with contracts, the majority lived in Petare, where three out of five had contracts. In the central areas four out of five had no kind of written contract. The vast majority of tenants (81%) had put down a deposit, and a further 6% paid in advance.

Despite the landlords' preference for the rent to be paid in advance, less than one-third of tenants did so. However, non-payment does not seem to be a great problem. Only one in six tenants claimed to have delayed paying the rent in the previous year, and in most cases this was not for more than 1 or 2 months. The possession of animals and/or children was the most common reason why landlords rejected applicants. One-fifth did not want animals and almost two-thirds mentioned some combination of not wanting animals or children. Two-fifths of landlords demanded an advance payment as the main form of security, while almost half wanted the rent paid in advance. Guarantors were almost never used.

Evictions occur, but among the tenants who had left a previous rental home only one in five mentioned that the family had been evicted. Certainly, the average tenure of 6 years in the current house suggests that eviction is not a major problem. There seems to be little variation between the central areas and the consolidated settlements in terms of tenure, although the rate of eviction in San José was higher.

On the whole, relations between landlords and tenants seem to be benign; neither landlords nor tenants complained much about the other. Few landlords claimed to have had problems with their tenants and, of the handful who did complain, drunkenness and noise were the principal problems. Only two had taken any official action against tenants, and only 10% claimed to have evicted a tenant during the previous 12 months.

While tenants were more likely to complain, particularly about the accommodation or the quality of the services, three out of five said they had no problems, and in the central city the proportion was almost 7 out of 10. Complaints about living conditions were most common in the consolidated settlements. Relatively few, however, said that they had any problem with the landlord; the highest proportion being 14% in El Carmen de la Vega. Even when they were dissatisfied few tenants ever approached the authorities. Only 3 out of 276 tenants had approached a lawyer, only one had consulted a tenant or neighbourhood association. In fact, the latter kind of organization seems to play little rôle landlord–tenant relations, only two tenants admitting to be members of such an association.[1]

Rather than landlords and tenants living in conflict, the relationship between them can more often be described as friendly. Good relations are no doubt helped by the selection procedure, few landlords accept tenants who have not been recommended to them, but even some of the more commercially minded landlords seem to maintain decent relationships with their tenants. Civility is helped by physical proximity, most landlords and tenants live in the same settlement, and by the fact that most landlords and tenants have similar levels of income and kinds of social background. It is also helped by the long average rental tenure; over the years landlord and tenant get to know one another quite well.

Only in the central areas are landlord–tenant relations more difficult. Certainly, a higher rate of eviction, at least in San José, does not help good relationships. Nor do many tenants in the central area know their landlords personally. Selection procedures are clearly rather different in the central zone, a factor linked no doubt to the high proportion of foreigners living there.

Conclusions

Rental housing is not the poor household's only housing option in Caracas. Self-help settlements offer an alternative for the very poor. Because poor families can obtain land free in an invasion settlement or, most commonly, purchase land or a shack cheaply from a previous invader, there is an alternative to rental accommodation. This is important in so far as rents are not cheap and at the time of the survey accounted for roughly 40% of the minimum wage.

Those who can afford to pay the rent can take advantage of the benefits that renting offers; notably the possibility of living in solid accommodation with services. That the option of renting is not ideal is shown by the fact that most tenants would like to be owners. But the advantages of renting are increased by the fact that landlord–tenant relations seem to be broadly amicable. There is no evidence of widespread abuse of tenants. Renting is an option that is particularly favourable for the many immigrants living in Caracas, many of whom no doubt harbour a desire to return home one day. While owner-occupation is open to them, they seem both more reluctant to take up this option and more able to pay the cost of rental accommodation.

It is this general housing situation which explains why most tenants are more affluent than owners in the new periphery. Households who move to very poor communities such as El Chorrito are being forced to choose

---

1 In fact, no tenant organizations appear to be functioning in Caracas.

the worst housing option available. They must live in flimsy shelters with minimal services. This is not the kind of ownership that they really want. Of course, if they persevere and work hard on consolidating their homes, if they continue in decently paid employment and are not sick, they will one day live in a consolidated and fully serviced settlement. And, as their children get older and begin to contribute something to the household budget, they will be able to expand and improve their property. It is unlikely that they will ever move house because very few owners in Caracas seem to move.

Eventually, some of these owners will begin to rent out accommodation, particularly those whose grown-up children have begun to leave home. These families will have made the slow and difficult transition from tenants to self-help landlords. Such owners will no doubt reflect on the paradox of their housing situation. As former self-help builders, who once could not afford to pay for rental accommodation, they will be providing rooms for those who are more affluent than they once were.

CHAPTER NINE

Drawing comparisons

Tenure trends

In all three cities there has been a consistent decline in recent decades in the proportion of families who are living in rental tenure. In Caracas, the proportion of tenants fell from 55% in 1961 to 30% in 1981, the proportion of owner households rising from 45% to 64%. In Mexico City, the proportion of households sharing or renting accommodation fell from 77% in 1960 to 46% in 1980. In Santiago, the share of tenant households fell from 57% in 1952 to 20% in 1982. Although there are signs that this trend may have been modified during the 1980s, the census figures are not yet available to support that contention.[1]

Despite similar patterns of relative decline, however, there are clear differences between the cities. In Caracas and Santiago, the absolute number of tenant households has remained more or less constant, whereas in Mexico City the number has increased dramatically. In the metropolitan area of Caracas, the number of tenant households rose from 155,000 to 176,000 between 1971 and 1981; in Santiago it fell from 220,000 to 197,000 between 1960 and 1982. In Mexico City, by contrast, the number of households in non-ownership rose 1.5 times between 1950 and 1980. The difference between the cities is partially explained by the varying rise in shared accommodation. This is especially important in Santiago, where approximately one-fifth of the city's families were thought to be sharing accommodation in 1983.

There has been a clear shift over time in the location of the rental

---

1 1990 data for the Federal District of Mexico suggest that the trend towards owner-occupation accelerated significantly during the 1980s. However, the change was most marked in the central area of the city. Since it was this area that was most profoundly affected by the 1985 earthquakes and subsequent government renewal policies, this may not be an accurate picture of the general trend in the city. Data for the State of Mexico are still not available.

housing stock. While the central areas remain important, the proportion of rental households near to the centre of the city has declined. This has been the result of urban decay, office and business development, and urban renewal programmes. The trend was hastened by the severe earthquakes that hit Mexico City and Santiago in 1985. Government policy has helped to redevelop these areas and reduce the numbers of tenants.

In recent decades, most of the new rental and shared accommodation has been created in the consolidated periphery. Whereas non-ownership in the central area of Mexico City grew by 10% between 1950 and 1980, in the rest of the city it expanded eightfold. In Santiago, while the central *conventillos* have been in decline there has been a clear process of densification in the consolidated suburbs; an outcome of the creation of new accommodation for both renters and sharers.

Despite the creation of this new rental accommodation, there has been a strong shift towards owner-occupation, a shift brought about by a variety of processes. First, the public sector has failed to produce housing for rent, turning increasingly to the construction of housing for owner-occupation. Secondly, there has been a clear decline, relative to demand, in the availability of rental housing in the central city; and in the absence of accommodation in their preferred location, many tenants have forsaken this form of tenure. Thirdly, improvements in transportation and infrastructure provision in the urban periphery have encouraged the expansion of owner-occupation. Fourthly, the failure of governments to control irregular forms of land occupation and subdivision has permitted many families to occupy land cheaply, sometimes even without charge. While this strategy changed markedly in Santiago under the military dictatorship, by 1973 a major shift in the tenure structure of that city had already occurred. Finally, ideology, cultural expectations and financial advantage combined after 1950 to convince most families that owner-occupation was the most desirable form of housing provision.

Only during the 1970s, in Santiago, and the 1980s, in Caracas and Mexico City, have there been some signs of change in this pattern. A severe decline in household incomes has affected families in all three cities. This has made home ownership a more problematic option for the poor and the middle class alike. In addition, there has been an increasing shortage of land. In Caracas and Mexico City, the authorities have made a partial attempt to increase residential densities and to control urban sprawl, but the principal factor has been the increasing distance between the new housing areas and centres of employment. In Santiago, the constraint has been directly due to government policy; irregular forms of land acquisition have been prohibited. The result may be a rise in rental and shared housing.

Differences between owners, tenants and sharers

Many international studies show that owners differ significantly from tenants in terms of their socio-economic characteristics (Downs 1983, Harloe 1985). This study placed considerable emphasis on attempting to discover whether there were systematic variations between tenure groups in Caracas, Mexico City and Santiago.

The basic finding must be that in none of the three cities can owners, tenants and sharers be separated into clear socio-economic groupings. There were too many similarities between owners and tenants, and even between tenants and landlords. Owners, tenants and sharers do not constitute homogeneous groups, and no single factor seems to determine whether households are tenants, sharers or owners. Nevertheless, some significant patterns were discovered.

Income and tenure

Many previous studies of tenure have shown that owner households tend to be more affluent than tenant households, although the differences have been less obvious in Latin American cities (Edwards 1982, Gilbert 1983, Gilbert & Varley 1991). Differences between owners and tenants are sometimes due to higher male incomes but more often because of the larger contribution made by other family members in owner households. The findings from this study differed in certain respects from this pattern.

In Santiago and Mexico City, the incomes of owner households were higher than those of non-owners, although in Caracas tenants earned more than owner households. When per capita incomes are considered, however, tenants proved generally to be better off than owners in all three cities, a result of having smaller families. In Caracas and Mexico City, there were considerable variations in the incomes of owner households, with new owners comparing very badly with owners in the consolidated settlements. Indeed, new owners in both cities were poorer than any group of tenants in terms of both household and per capita incomes. In Santiago, of course, new owners were drawn from a very different income group because of the system used to select families in official settlements.

Among the tenants, those in the better locations tended to be the most affluent, although some poor tenants were found in every settlement. Those living in the central city had higher per capita incomes in Caracas and Santiago. In Mexico City, however, tenants in one highly consolidated settlement were a little better off than the central tenants, although the latter were still better off than any group of owners. The tenants were generally the most affluent of the non-owning groups.

Sharers in Mexico City had lower household incomes than other tenure groups but were more affluent than many of the owners in per capita terms. In Santiago, sharers were generally poorer than either owners or tenants.

The similarities in household income were closely reflected in the ownership of consumer durables. In all three cities, there were few differences between owners and non-owners, although new owners had fewer refrigerators, washing machines and cars.

In Caracas and Mexico City, therefore, there is evidence that new owners are among the poorest households. This finding suggests that this group may have been forced into owner-occupation by their low incomes. This is most obviously true of owners living in invasion settlements who were able to obtain land at no cost, or at least very cheaply. Certainly, the owners in the new invasion settlements in Caracas were very poor, and in Mexico City new owners in the invasion settlement were poorer than those in the illegal subdivision. A similar pattern is also borne out by the Santiago data in the case of the settlement founded by invasion in the early 1970s. More recently, of course, the link between poverty and owner-occupation has been reversed in Santiago. Now, the local team claims, "families that cannot meet the rules of eligibility for subsidized homes are condemned to maintain themselves as tenants and pay a higher proportion of their income for housing.".

Age and tenure

Tenants and sharers tend to be drawn from a younger age group than owners, but again there were exceptions. In Caracas, for example, there was no overall pattern; while tenants were generally much younger than owners in the consolidated settlements, owners in the new settlements were the same age as tenants in the central city. In Mexico City, however, most sharers and tenants were much younger than owners, the only exception being in the central areas, where tenants were older than new owners although younger than owners in the consolidated periphery. The comparative youth of the tenants and sharers was linked to the fact that this was the most common tenure form among newly formed households. In Santiago, sharers constituted the youngest group and so-called "new" owners were comparatively old because of the prohibition on land invasions in recent years. What is particularly interesting in Santiago is the difference between the age of adult sharers, 31 years, and that of the allegados, 55 years. In practice, the age pattern in Santiago was complicated by the very different ages of household heads in each survey settlement.

Family structure and tenure

Owners had larger households than non-owners in all three cities. The nuclear family was larger because owners generally had more children. In Caracas and Santiago, owner households contained more extended families, although in Mexico City there seemed to be little difference. In new settlements most owner households contained adult males, but in the older settlements there was a much higher proportion of widows and

142

separated women. In one consolidated settlement in Santiago, 38% of owner households lacked an adult male.

Among the tenants, centrally located households tended to be smaller than those elsewhere. Certainly, in Caracas and Mexico City, central tenants had many fewer children, and single-headed households were common. Sharer households tended to be nuclear in form and most contained males. In Santiago, however, *allegado* households frequently lacked an adult male or female.

Generally, the size of the family seems to be an important influence on residential tenure, although the Mexico study concluded that "having children (rather than the number) is what modifies the needs and expectations of most couples.". Having the first child seems to be more significant than the birth of subsequent children. Having children affects housing behaviour, once you have a couple it seems another one doesn't make much difference. The effect is less due to demands for space than the widespread feeling that families wish to have something to leave to the children. In Santiago, the number of children seemed to make little difference to tenure, with most families have grown in the existing house. The birth of additional children seemed not to trigger house moves, an effect perhaps of the very long stays typical of most families in the city.

Employment and tenure

Earlier work in Mexico had shown that owners were more likely to be drawn from the self-employed and to have had some experience in the construction industry (Gilbert & Varley 1991). But in Caracas there seemed to be few clear links between employment and tenure. The nature of work seemed to be affected more by the location of each settlement. Similarly, in Mexico City no clear pattern emerged, although among owners there was a higher proportion of self-employment and of construction workers. In Santiago, the significant age differences between the settlements meant that little meaning could be read into the employment structures.

Migration and tenure

Previous work in Guadalajara and Puebla had discovered a much higher incidence of migrants among owner-occupiers. It was not surprising, therefore, that the number of migrant households found in the new peripheral settlements of Mexico City was much higher than their relative weight in the total sample. By contrast, tenants, especially in the central city, were much more likely to be natives. It seems that native Mexicans use their family networks to improve their housing situation, either through the offer of accommodation or through loans with which to buy better-quality plots or homes in the periphery. Given fewer alternatives, migrants are often obliged to move into lower-quality owner-occupation. There were also some intriguing correlations between birthplace and tenure in Caracas, although here there was a complicating feature, the fact that so many of the migrants were actually foreign-born. The immigrants were heavily concentrated in the central areas and did not often become owners. Among the Venezuelans, however, there appeared to be only a weak link between tenure and place of birth. Owners were more likely to be natives but migrants were extensively represented both among owners and tenants. In Santiago, there was some tendency for natives to out-number migrants among the non-owners but the difference was not great. Where place of birth did seem to be influential was among the *allegados*. Here there was a much higher proportion of natives; a clear outcome of this group having more contacts and therefore possibilities to share accommodation.

It is clear that the structure of the housing market in each of the three cities is sufficiently different to complicate any simple pattern between socio-economic characteristics and tenure. Age, family structure, income and migrant status all contribute something to the explanation, but none correlate very closely. A more complex pattern is also encouraged by the diversity that exists within each tenure group. Indeed, it is clear that only if we examine the characteristics of each subgroup, can we make real sense of their residential behaviour.

Housing conditions

Housing conditions vary considerably between and within tenure groups. In general, however, owners in consolidated settlements occupy well-constructed homes which have been fully serviced. Of course, this was not always the case. Most have lived a long time in the settlement and many lived for a number of years in poor conditions. Only now have circumstances improved. By contrast, new owners in Caracas and Mexico City are at the beginning of the self-help consolidation process and are currently living in very poor conditions. The situation is different in Santiago because there are no new self-help settlements.

Tenants generally occupy less space than owners. However, the quality of that accommodation is superior; tenants only rent property that is supplied with infrastructure and services, they do not rent accommodation that is flimsy. Therefore, tenants gather in accommodation that is well established, and the proportion of tenants rises with the age of a self-help settlement (Gilbert & Ward 1985, van Lindert 1991). Crowding tends to be highest in the central city, but the tenants benefit from the location and often from superior services.

Sharers also live in somewhat cramped conditions and the quality of the accommodation tends to be lower than that of tenants. Their shelter is much better, however, than that of the new owners.

Preference for ownership

Respondents in all three cities expressed a clear preference for ownership. The vast majority of owners were content with their current tenure, practically all having had previous experience as tenants or sharers.

Among the tenants, the majority wished to be owners. In Santiago, 86% of tenants said that they would prefer to be owners and in Caracas, 94%. Only in Mexico City was the preference less marked, some 58% saying that they would prefer to own. Patterns of residential movement in each city reflected this general preference for ownership. There were very few cases of families now renting or sharing accommodation who had previously been owners.

When owners were asked about the advantages of owning a home, one general sentiment came through above all else. Ownership gave a sense of security and a feeling of independence. Having something of one's own gave an important boost to a family's self-esteem. Ownership also offered some direct advantages. It freed households from the need to pay rent, it meant that they were more independent from kin, it gave them something that they could leave to the children. In general, it increased their "peace of mind" (tranquilidad).

Such an attitude can only be understood in the context of the political economy of each city. Many of the perceived advantages of ownership have been created quite deliberately by state action. This is most obvious in Santiago, where successive governments have offered home owners generous subsidies. As the Santiago team put it: "the treatment accorded by the state to owners and tenants is very different and highly regressive with respect to tenants.". In Caracas, ownership has been facilitated among the poor by the lax attitude of the authorities to land invasion. Self-help ownership has been permitted through the occupation of state land. In Mexico City, land has rarely been so freely available, although peripheral plots have still been cheap. There the authorities have also turned a blind eye to illegal and unserviced subdivision.

Despite this general preference for ownership, some households remain in rental or shared accommodation even though they have the resources to acquire their own self-help home. Tenants living in crowded conditions, who could have had much more space as home owners on the periphery, clung to their rental accommodation. Among these tenants there was a substantial number who dissented from the chorus of approval for ownership. In Mexico City, two-fifths of central tenants preferred to rent, and in Santiago, 30% of those living in the *conventillos* said the same. Four-fifths of these central tenants had never looked for their own home.

There are a number of explanations of this seemingly irrational attachment to tenancy. First, renting in all three cities often matches ownership in terms of security of tenure. Tenants are not constantly fighting against eviction. Few tenants change home frequently, indeed, in the central areas of Santiago and Mexico City, households stay in the same house for a very long time. If security of tenure is supposedly one of the main advantages of ownership, many families in Mexico City and Santiago seem to achieve this goal perfectly satisfactorily through renting.

Secondly, living conditions for many tenant families are superior to those of owners. Because of their age, most buildings in the central city and in the consolidated settlements are provided with water, electricity and drainage. By contrast, the homes of the new owners lack many of these essential services. But, it is not only the loss of services that is important to tenants, it is also the loss of an established community. Many sharers and tenants feel part of the community in which they now live. A move to the distant periphery would disrupt their existing social networks. In Caracas, feeling part of a *barrio* community was a frequently expressed sentiment underlying the reluctance to move into owner-occupation.

Thirdly, location seems to be a critical factor influencing the desirability or otherwise of the ownership option. Whereas tenants in the central city live close to work, and those in the consolidated settlements have access to relatively good public transport, journeys from the new periphery can be both long and arduous. To travel to the centre from the new settlements of Caracas or Mexico City is often a very tiring and frustrating experience.[1]

Fourthly, it is quite clear that self-help construction does not represent an easy route into ownership. It costs a great deal of effort in terms of the labour that must be put into the construction process and involves suffering poor living conditions over a number of years. As the Mexico team describe the choice: "the great majority aspire to owner-occupation but not under the conditions that are implicit in the ownership and self-help construction of a house in the periphery.". Some families will decide that they are unwilling and/or unable to face this burden.

Fifthly, some families are in a better position to build in the periphery. Certainly, when we look at the characteristics of families who do become owner-occupiers, certain features stand out. Families capable of under-taking the task of self-help construction are found in much greater numbers among the owners than among the tenants. Thus the proportion of workers in the construction industry is usually higher among the owner households, and the number of female-headed households much lower.

Sixthly, not only are some families in a better position to build in the periphery, but others are in a better position to avoid the need to do so. This seems to be illustrated very well by the differences in the location of native and migrant households in Mexico City. Whereas a relatively high

---

1 In a city such as La Paz, where most of the new self-help settlements are located in harsh physical areas, location becomes still more significant (van Lindert 1991).

proportion of native families live in the central city, migrant households are more likely to be found in the distant periphery. The former have a wider range of accommodation options. In addition, when natives obtain plots they tend to be in better locations or are better serviced than the plots acquired by migrants. They can make use of the "waiting space" offered by their ability to share with parents or kin.[1]

Finally, it is clear that the rental alternative is not equally attractive in all three cities. It makes more sense to rent in some cities than in others. Renting a home in Mexico City is much cheaper than renting a home in Santiago or Caracas. There are relatively few families in Santiago who can afford to rent accommodation, whereas rent–income ratios in Mexico City are much lower. Under these circumstances fewer families in Mexico will be forced into peripheral ownership by their inability to pay rent. Families in Mexico City face a much more balanced choice between renting and owning than in Caracas or Santiago. Conditions may not be ideal in the rental accommodation, but it does offer specific advantages at a reasonable price. In Santiago, renting is simply unaffordable for many families, in Caracas, rents can be paid only with difficulty.

What these findings suggest, therefore, is that there is a significant difference between the general desire for ownership and the practicalities of becoming an owner. When tenants say that they want their own homes, they are expressing their preference for a particular form of ownership. They are saying that they want to own a particular kind of home, maybe one similar to the home that they are now currently renting or maybe even a palace. If what the market offers is not this kind of home, some households may choose to forego the opportunity for owner- ship until their preferred kind of home becomes available. Ownership remains an aspiration which will only be taken up when they find the kind of ownership option that they want. Meanwhile others who may harbour a much less overt desire for ownership may find themselves in circumstances that either encourage purchase or which may push them into it. If they cannot afford to pay the rent, then they may have to find another form of tenure.

In sum, the goal of home ownership is much cherished in all three cities. However, the form in which home ownership becomes available is highly significant. Some families will accept home ownership even when it means building a home on the unserviced periphery, for others this is anathema. For some, ownership on the periphery is highly desirable but is an unobtainable dream. The result is that we have a highly diverse response to seemingly the same aspiration.

---

1 See Gilbert & Varley (1991) for similar findings for Guadalajara and Puebla, and van Lindert (1991) in La Paz.

Sharing versus renting

Why do some households rent and others share? Clearly, some households have no choice, for those with very low incomes, sharing with kin may be the only real alternative. This is clearly the case with many *allegado* families in Santiago. The high cost prevents them from renting and the inability to find land for self-help housing precludes ownership. Family circumstances, low incomes and incomplete households add to the inevitability of their choice.

Of course, while desperate families may be forced into sharing, not every household has kin or friends in the city able to accommodate them. For such families, the option of sharing is simply not available. For this reason, it is clear that migrants are less likely to share accommodation than natives.

But, too much should not be made of the issue of compulsion. For, if some sharers are forced into their present form of tenure, this does not mean that all are unhappy with it. Indeed, most Mexican sharers prefer to live with kin than to rent accommodation. They do not have to pay rent, they have access to the wide range of consumer durables belonging to their parents, and they have as much or more space than most tenants. Their only real complaint is about their lack of independence and ownership. Similarly, in Santiago, few who share wish to rent. Sharing is cheap whereas renting is expensive. Among the tenants there are some who would prefer to share were it possible.

Whether, of course, all of the accommodating households are happy with sharing is less certain. Some undoubtedly welcome having their children "at home" but others no doubt regret the additional pressure on space. Much will depend, no doubt, on the circumstances of the hosts. If they have a large house or plot, they may be pleased to accommodate kin; where they have little room their view may be more ambivalent.[1]

Residential movement

One of the most intriguing findings of the study is how static the housing market appears to be. Neither owners nor non-owners move very frequently. Once owners acquire a self-help home they seem not to move. This tendency is most marked in Santiago, where the mean length of residence in the consolidated periphery was 23 years, with most of the families having lived in the same house since they first became home

---

1 Clearly, size of plot has some influence here. In Bogotá, where plots are relatively small, there certainly seems to be less sharing than in Mexico City, where many plots are quite large (see Gilbert & Ward 1985).

owners. In Mexico City, owners average 14 years in their homes in the consolidated settlements; and in Caracas, 18 years. Among the owners, the degree of stability seems almost to be a problem. Do they stay in their present home for so long because they are happy there, or is it because they simply cannot sell it?[1]

Tenants, especially in the central city, also tend to live for long periods in the same accommodation. This is particularly true in Mexico City and Santiago, where rent controls have helped some families to retain their accommodation over years and even over generations. In Santiago, one-quarter of the tenants in the *conventillos* had lived more than 40 years in the current home; in Mexico City, almost three-quarters had lived for more than ten years in the same home. A different pattern was apparent among the central tenants of Caracas, the average tenure being only five years. Clearly, the urban structure of Caracas is different, rent controls have been less influential, and the widespread presence of immigrants also affects the length of tenure in the centre. Nevertheless, tenancies are hardly short.

Tenants in the periphery also live reasonably stable lives. In Mexico City and Santiago, the average tenure is three years, and in Caracas seven years. Eviction is a worry for the tenant population, as are increases in rent, but these perennial problems do not seem to have produced the instability that was so characteristic of, say, 19th-century British cities (Englander 1983, Kemp 1987).

When households do move they tend not to move far. Most tenants move less than 5km, and many owners move similarly short distances. While the owners move farther they rarely cross from one side of the city to the other, the exception being Santiago where compulsory relocation forced many owners to do just that. In Caracas, only one-third of owners had moved more than 5km in their last moves, and virtually none had moved from the east of the city to the west or vice versa. In Mexico City, the pattern of movement was affected by the date of formation of the settlement and the form of land occupation. All owners moved within the same sector of the city, but whereas around half of those in the consolidated periphery had lived previously within 5km of their present home, in the new settlements a majority of owners had moved at least 10km. However, the way a settlement was founded also seems to make a difference. There was also a difference according to the formation of the settlements; a much higher proportion of the population came from nearby communities in the invasion settlements than was the case in the illegal subdivisions.

---

1 In fact, we know rather little about the buying and selling of self-help homes, or for that matter any homes in Latin American cities. Clearly, disposing of a property can be difficult given the lack of title deeds and/or the lack of credit facilities for the purchase of second-hand homes.

The production of rental housing

Landlords are older than other owners and much older than most tenants. They have smaller families, are more likely to be divorced or single and are much more likely to be retired. Landlords are also more likely to be self-employed. They live in larger properties than other families and have lived longer in their current home. In Caracas and Mexico City, their household incomes are higher than those of other owners, although this is not the case in Santiago, but in all three cities their per capita incomes are higher. While they are generally more affluent than other owners, they have similar per capita incomes to those of their tenants. Clearly, landlords, owners and tenants in the consolidated settlements are drawn from the same social class.

Landlords vary from the large-scale owner of central property to the landlord with a single sitting tenant. In the central city, both large-scale and small-scale landlords can be found, some operating on a wholly commercial basis while others are merely scraping together some kind of income from their only asset, their property. In Santiago, many tenants sublet property because the landlords do not want the problems of dealing with many tenants. Some land is also sublet to poor families so that they can build their own shelter. In Mexico City, many of the larger operators employ administrative agencies who represent the visible face of the landlord. These companies not only let and administer the property but buy and sell it. There is no longer much concentration of ownership, inheritance has helped to break it up. In the central areas it is certainly more economic to sell property than to rent it, but the landlords complain that this option is often made difficult by the effects of rental legislation and the actions of tenant associations.

In the consolidated periphery, most landlords operate on a small scale. In Santiago, 7 out of 10 landlords only let to one tenant household, in Mexico City three-quarters, and in Caracas two-thirds. Few landlords had been renting for very long; in Santiago only 45% had been in business for five years or more, in Mexico City two-thirds had been letting property for three years or less.

While some landlords had constructed separate dwellings with the intention of renting them out, most landlords let rooms in their own home. For this latter group, entry into landlordism is often stimulated by the departure of grown-up children. Spare rooms are converted into a source of income. Such landlords may even evict the tenants in the event that their children wish to return home. In Mexico City, landlords tend to live separately, unless they rent in invasion settlements where they fear that in their absence the tenants may take over their property. In Santiago, many landlords merely let the land, the tenant building a separate dwelling at the back of the owner's plot. This phenomenon became widespread during the military dictatorship when the Church

150

introduced its Home of Christ (*Hogar de Cristo*) programme.

The research shows, therefore, that few landlords follow a commercial rationale. Indeed, it is this lack of capitalist behaviour that sustains the expansion of rental housing. Many of the landlords recognize that renting does not produce a large income. In Caracas and Santiago, few "domestic" landlords thought that they got much out of letting, especially given the hassle involved. In Mexico, the reaction was more variable, with some small-scale landlords admitting that it was a satisfactory business. In no sense, however, are most of these landlords operating on a commercial basis. They are merely supplementing their income, an income severely reduced by the effects of inflation, and trying to provide themselves and their children with a slightly safer financial future. They do not know how to invest in, nor do they really trust, other forms of investment, therefore their spare money goes into housing and land. This is a very different landlord from that depicted in writings about 19th-century Europe.

Relations between landlords and tenants

Relations between landlords and tenants are frequently portrayed as posing one of the major difficulties of rental accommodation. Landlords denounce the actions of tenants, tenants those of landlords; the frequency of confrontation is emphasized by both sides. Our research shows that relations between landlords and tenants is much more benign, even amicable, than that portrayed by this picture. Relatively few landlords and tenants speak badly of one another.

The most common feature of landlord–tenant relations in the consolidated settlements is the lax legal basis and the generally friendly relationship. Good relations are helped by the selection procedure, few landlords accept tenants who they do not know or, most typically, who have not been recommended to them by friends. In Mexico City, one-third of the tenants already knew their landlord when they first rented the accommodation, and many of the others were recommended to the landlord through friends or kin. Between one-fifth and one-quarter of tenants maintained active social relations with their landlord. Civility is important because both landlords and tenants live in the same settlement, and contact is eased by the fact that most landlords and their tenants had similar kinds of incomes and social backgrounds. It was also helped by the long average rental tenure, over the years landlord and tenant got to know one another quite well.

Bad landlord–tenant relations are more common in the central areas. In central Caracas, housing conditions are certainly worse, relations less friendly and evictions more common. This situation is accentuated by the fact that one-quarter or more of the tenants are foreign and many fewer landlords live either in the property or in the neighbourhood. Indeed the majority of tenants do not know their landlord, a tendency accentuated by relatively short stays. In Mexico City, few tenants in the central areas know their landlord and most are issued with formal contracts. Conflict is much more common in the central areas and in all three cities there are many more complaints from central tenants about deficient services and poor housing conditions.

One way of measuring the degree of harmony between landlords and tenants is through the frequency of evictions. Of course, eviction is by no means uncommon in any of the cities. In Caracas, 12% of tenants had left their penultimate home through eviction; in Mexico City, 20% of tenants had left their previous home through eviction and another 13% had left because the owner had sold the property; and, in Santiago, 28% of tenant families with a previous rental home had been evicted. These figures do not include cases where tenants left because the rent was raised, a case which, the Mexico team argues, constitutes *de facto* eviction. While their point is often valid, inclusion of all such cases would surely overestimate the real number of evictions.

However, eviction cannot be such a significant issue in cities where so many tenants have long tenures. In Caracas, more than one-quarter of tenants had lived for more than five years in the previous home, and more than half had rented only one home in the previous five years. In Mexico City, the average tenancy was seven years in the current home and six years in the previous rented accommodation. In Santiago, tenants had spent an average of six years in the current home.

In the central areas, tenancies are usually even longer. In Mexico City, the average tenancy in the previous two homes was 17 years and 8 years, respectively, and, in Santiago, 46% of central tenants had lived for more than ten years in the current home. Only in Caracas were central tenancies much shorter, something clearly linked to the high proportion of immigrants among the tenants.

Political organization

In the consolidated settlements, there was relatively little sign of organization on the part of either landlords or tenants. In Santiago, one-fifth of tenants belonged to some kind of neighbourhood association, but none of these organizations primarily represented tenants. As a result, no resident landlord could recall any problem caused by such an association. The picture was similar in Caracas and Mexico City.

Only in the central areas were tenant organizations more common. In the central areas of Mexico City, 9 out of 80 tenants interviewed belonged to a tenant organization and, in Santiago, tenant organizations were sometimes active in opposing evictions.

Towards an interpretation

Early work on residential behaviour in Latin American cities emphasized the relationship between tenure and migrant status (Turner 1967, 1968). The bridgeheader/consolidator model, devised more than two decades ago, still has a certain validity. Migrants still follow the transition from renting to ownership. There is a pattern of movement from more central areas outwards towards the periphery. Poor households are able to build and consolidate homes, at least when they can create an investment surplus. But, now that cityward migration has been under way for so many years, more and more people have been born in the city. As the cities have become larger the housing alternatives become more varied and complex to describe (Brown & Conway 1980, Gilbert & Ward 1982, Gilbert & Varley 1991, van Lindert 1991).

The importance of city size is shown particularly clearly by housing behaviour in Mexico City. Here there are many different forms of accommodation, varying in terms of location, tenure and quality. The poor in Mexico City are accommodated in a wide range of housing forms: renting in the deteriorated central city; renting in the consolidated periphery; renting plots in "lost cities" or rooms on rooftops; the purchase of subsidized social-interest housing in the periphery; the purchase of renovated housing in the city centre; self-help construction in the consolidated periphery; self-help construction in the new periphery; and sharing a house or lot with kin or friends. While none of these options is equally open to every family, there is a much wider range of alternatives than is available in most smaller cities.

What this variety strongly underlines, therefore, is that in any residential model we need to consider carefully the options available to the poor in their housing choice. To consider the demand side of the tenure equation is meaningless without understanding the range of options available to poor families. No choice is without constraint, we have to consider barriers as well as opportunities.

It is clear that the only way in which poor families can attain owner-occupation in most Latin American cities is through self-help construction. Given the distribution of income, the nature of the land market and the structure of the building industry, self-help housing is the only real alternative. But ownership of a self-help home is only possible where land can be obtained. The basic ingredient in understanding the transition to ownership in Latin American cities, therefore, is to consider carefully how readily poor families can obtain land.

In practice, there is a great deal of variation. Certainly, in some Latin American cities, land is both relatively cheap and accessible. In the Lima that Turner described in the 1950s, and on which he based his model, land invasions were possible on land that was distant but not excessively far from the main areas of work (Collier 1976, Riofrio 1978, Dietz 1981).

That is no longer the case in Caracas, Mexico City or Santiago. In all three cities there are major barriers to the poor acquiring cheap land. In Caracas, located as it is in a series of narrow valleys, land is now extremely scarce; free lots are now found only in areas distant from the city centre. In Santiago, invasions used to be the main source of land but under the Pinochet regime *tomas* were prohibited. In Mexico City, irregular forms of subdivision provide the poor with land that is still quite cheap relative to incomes, even if it lacks services and is available only in distant locations.

If access to serviced land is effectively denied to most of the poor in all of these cities, there are none the less important differences between them. In Santiago, there is no self-help land of any kind, while in Caracas distant land is freely available and serviced plots are on offer at a price in the more consolidated parts of the city. In Mexico City, cheap land is available in the periphery and some relatively expensive plots in the consolidated self-help settlements.

These differences mean that the desperately poor in Caracas and Mexico City have more housing options than in Santiago. They can rent or share, or they can invade or occupy cheaply land in the distant periphery. Living conditions are rudimentary but it is still an option that is not available to the poor of Santiago. In the latter, the high cost of land means that a higher income is necessary to gain access to home ownership; a barrier only partly mitigated by generous housing subsidies.

The availability of cheap plots is not only significant in determining the numbers of poor households gaining access to land. It also affects the level of rents, which differ considerably between cities. It is clear that families in Mexico City pay much less in rent than families in Santiago. The higher prices of land in Santiago have had the effect of raising overall rent levels. Since it is financially more difficult to move into home ownership, there is a knock-on effect on rents.

Of course, the cost and form of land access is not the only influence on rents. Government legislation also has some effect, for example, in holding down the rents for some tenants in the centre of Mexico City. Similarly, different patterns of social relations between landlords and tenants have a considerable effect on rents. The fact that Mexican tenants stay in their rented accommodation for long periods helps to lower Mexican rent levels (Gilbert & Varley 1991). Since landlords admit to raising rents much more for new tenants than for sitting tenants, a low turnover rate has the effect of holding down rent levels. It should also be remembered that rents vary quite dramatically through time, even within the same city. In Mexico, rent levels fell by half relative to the general price index between July 1970 and July 1987; during 1988 and 1989, however, they rose more than twice as fast.

The importance of these variations in rents and purchase costs is that they are likely to influence the desirability of ownership *vis-à-vis* renting.

Clearly when rents are very low, families will continue in rental accommodation even when the accommodation is inadequate. Should rents rise, they may well reconsider their housing situation, either because they cannot afford the higher rent or because the balance of advantage between ownership and renting has shifted. It is this balance of advantage, not just in costs but in convenience, servicing and location, that seems to be critical in the process of residential choice. This balance, of course, is not determined by individual families but by the political economy of land and housing in the city and country concerned.

It is for these reasons that the results of this and previous research do not produce a simple explanation of low-income housing behaviour. Not all poor families rent, not all poor families occupy shanty towns. Within cities there is a diversity of response; between cities still greater variation. As such it is erroneous to make general statements either about the desirability of renting *vis-à-vis* ownership or the nature of tenants and owners. There are certain similarities but there will be major variations between cities according to the production system. Households have different needs but they are forced to modify their behaviour according to the different circumstances facing them. Thus, in Santiago, many households have been forced to double up with kin or friends because no cheap alternative is available. In Mexico City, we find single women building homes in invasion settlements because, despite the immense problems they face, land is free and no rent has to be paid for a self-help home. We also find that there is a higher incidence of extended families in the central slums than in the self-help periphery; location and services are more important than space to these families.

This diversity also explains why research in different cities comes to different conclusions about the nature of owners and tenants. Thus, some years ago on the basis of research in Bogotá, I concluded that renters and sharers "are an excluded majority, excluded from an alternative that few would regard as positively desirable" (Gilbert 1983). The research in Mexico City possibly leads to the opposite conclusion that "it isn't a high income that permits home ownership, but precisely the lack of one. It is economic pressure, the impossibility of paying a rent, the need to have 'some kind of roof in order to live an independent life' that leads them to obtain a lot and build their own home, notwithstanding the poor conditions and the lack of services in the settlement." There is no contradiction between these different conclusions; the explanation behind the difference is that the opportunities facing the poor household in search of a home vary considerably from city to city. Their socio-economic characteristics and tenure vary accordingly.

CHAPTER TEN

By way of conclusion

The results of this research project have made it abundantly clear that rental housing sometimes provides a thoroughly satisfactory housing solution. It is wrong to assume that renting is always a bad option for poor families. Although most poor families state a preference for home ownership, that is not necessarily an option that they wish to take up right now. Home ownership may be their ultimate dream but rental housing may be what they need today. For rental housing offers what home ownership rarely does in Latin American cities today, accommodation in convenient locations and the opportunity to change homes easily. Many families prefer to take advantage of these benefits even though they can afford to buy plots and build in the distant periphery.

Households adapt their aspirations and desires to the realities of the housing and land-supply systems. In understanding residential behaviour, therefore, we need to examine not only the relative advantages of ownership vis-à-vis those of renting, but the advantages and disadvantages offered by different kinds of rental and ownership package. Households clearly choose different tenures and forms of housing, according to their family circumstances. Whereas some are quite prepared to struggle through self-help building and consolidation to improve their housing situation, others decline that option. Some families are well able to build a home, others are very poorly equipped to do so. Renting is a more appropriate tenure for certain kinds of household: those that are mobile, those with workers employed in the central areas, new families, and households headed by either the old or the infirm.

By now, it should also be obvious that the periphery is no longer occupied exclusively by owner–occupiers. The creation of new settlements of self-help builders soon creates new opportunities for tenants. Given 3 or 4 years, a well-located settlement with transport facilities and services will have developed an extensive rental system. When we refer to irregular and self-help settlements, therefore, we should recall that in the more consolidated neighbourhoods there may well be more sharers and

156

tenants than owner–occupiers. It is also clear that, unlike the picture drawn by so many political groups, relations between most landlords and their tenants in these areas are amicable. While differences of interest are apparent at times, most landlords and tenants coexist reasonably well. Nor are rents always exorbitant, their level relative to income varies considerably between cities (Malpezzi & Mayo 1987, Gilbert & Varley 1991).

It is also clear from this research that while government policy influences the rental housing sector, its impact is very partial. Over the years, governments in Chile, Mexico and Venezuela have introduced legislation to regulate rental housing. They have attempted to control rents, landlord–tenant relations and environmental standards. What is also abundantly clear is that little of this legislation has much effect in the low-income suburbs. In Caracas, although legal powers of intervention have existed since 1921, the law has rarely been enforced. Few landlords outside the central area issue contracts, and even fewer notarize contracts. Similarly, although half of the tenants in the central areas of Mexico City have contracts, virtually no contracts are issued in the periphery. In Mexico, the landlords claim that the law has no real influence on their relationship with the tenants; certainly neither landlords nor tenants seemed very knowledgeable about the law. In the periphery of Santiago, the situation is very similar. Should governments wish to legislate, therefore, it is by no means certain that their laws will make much difference in the low-income settlements.

It is also salutary to recall that landlord–tenant relations seem to function better in the consolidated settlements of the city where the law does not apply. In the central areas, landlord–tenant relations are often much more conflictive. It cannot be assumed, therefore, that greater state intervention in landlord–tenant relations is helpful. Currently, in its critical rôle of providing some kind of arbitration mechanism between aggrieved parties, the legal system fails badly. In all three cities, the judicial system is both too slow and too expensive to afford much help to either landlords or tenants.

In addition, where government legislation has had an impact, the results have not always been those that had been anticipated. This is most clearly demonstrated by the effects of Mexican planning legislation in the 1940s which, by raising physical standards, increased the risks of prosecution for those building cheap *vecindades*. Since landlords have always faced some kind of trade-off between higher housing standards and profits, the introduction of more exacting standards encouraged many to direct their funds into some other kind of business. Indeed, in Mexico City the introduction of higher planning standards probably had more influence over the rental sector than the infamous policy of freezing rents. The latter damaged the rental housing sector by severely restricting residential movement; sitting tenants refused to move and their continued

157

presence made property difficult to sell. Such outcomes are not an argument against appropriate forms of state intervention and certainly not a plea for the instant removal of rent controls. They merely alert us to the need for a re-evaluation of existing policies, especially those that discourage the production of rental housing (Malpezzi & Ball 1991).

Households exhibit a wide range of housing needs and seek to improve their housing in different ways: "People may improve their shelter conditions by moving into ownership, or by moving into a better serviced neighbourhood, or by enhancing the standard of the occupied structure itself through self-built activities" (van Western & van Lindert 1991). As such, future government policy should seek to create a matching variety of housing options. Since rental accommodation offers the kind of shelter required by certain household types, particularly the young and the mobile, any city that lacks sufficient rental housing soon begins to regret it. To simply encourage owner-occupation in Latin American cities is neither feasible nor desirable. To ignore rental housing, given that up to half of the population is living in such accommodation, is simply irresponsible. Renting has to be recognized as both a respectable and a necessary housing option.

Because few Latin American governments seem to be prepared to act as social landlords, any attempt to increase rental housing production requires some form of encouragement for the private landlord. In so far as the state is prepared to give housing subsidies, balance is necessary. Subsidies should not be directed only to owner–occupiers, as is current government practice in Chile, Mexico and Venezuela; if subsidies are to be made available, private rental accommodation should be eligible for assistance.

In considering incentives for rental housing, governments should not be too concerned about the distributive effects of subsidies. Of course, in some places incentives to private landlords may worsen the distribution of income, but in the cities considered here this does not appear to be a real danger. Most landlords are small-scale operators and few are letting on a true business footing. In the consolidated self-help settlements, the socio-economic characteristics of landlords are rather similar to those of their tenants. Incentives to landlords, therefore, are unlikely to worsen the distribution of income.

When governments intervene, encouragement for rental housing has to be considered as part of a much wider social and economic strategy. Appropriate intervention may require change in a whole series of policy areas. It may require amendments to tax policy, changes in the system of allocating subsidies, the reform of judicial procedures, and perhaps the creation of special credit programmes for landlords. Even then, success depends on the nature of general housing policy, particularly the attitude towards home ownership and to irregular forms of land occupation. If cheap land is available on the edge of the city and services and legal titles are supplied quickly to self-help settlements, the demand for rental housing may well decline.

Under current circumstances, however, it seems appropriate to seek to encourage rental housing production. It also seems clear that any such encouragement should be given not to the formal-sector landlord but to the small-scale operator in the informal sector. Up to now, any subsidies have been given only to formal-sector building companies, so far with remarkably little success. Unfortunately, there is currently considerable reluctance to give help to the self-help landlord. In Mexico, the authorities are worried that by helping landlords in unregularized areas they will encourage the spread of irregular forms of settlement. And yet, it is clear that a modicum of intervention would yield a substantial increase in the number of rental housing units. Such intervention does not necessarily require financial incentives. A most effective incentive would be the offer of title deeds to owner–occupiers, freeing existing, and potential, landlords from the fear that ownership of their property will be claimed by the tenant. Equally helpful would be an improvement in services, infrastructure and transport provision, an approach that would help owners generally as well as attracting tenants into the upgraded settlements.

Of course, support for the expansion of rental housing in consolidated settlements is not without its dangers. Increasing population densities may stretch infrastructure capacity. Landlords may also extend their property in ways that could endanger life and limb. In Caracas, the instability of some of the hillside terrain is a particular problem should landlords decide to build a third, or even a fourth, storey onto their property. Clearly, some kinds of planning control are necessary.

If support for rental housing is appropriate in the periphery, something should also be done about rental accommodation in the central city. Centrally located accommodation is a necessity for many households. Moving to the distant periphery can cause them problems; it increases journey times and can easily weaken family-support systems. Some rental accommodation has to be available within easy range of the central city. Selling off rental accommodation to the existing tenants, as has been done in Mexico City as part of the post-earthquake rehabilitation programme, may be popular with the former tenants but blocks access to central accommodation for new cohorts of households wishing to rent.

There is, of course, little that can be done directly about other forms of non-ownership, such as sharing. In general, sharing is a direct reaction to housing shortage. Where families cannot afford to buy or rent the kind of shelter they desire, they take advantage of the offer made by friends or kin to share accommodation. This book has shown that the accommod-ation provided is sometimes superior to the alternatives available in other forms of tenure. But it is also clear that sharing is essentially a response to the lack of alternative forms of affordable accommodation. This is most

159

manifestly clear from the Chilean experience. By blocking all forms of irregular housing, the Pinochet government hugely expanded the numbers of *allegados*. While some forms of sharing are desirable, few Chileans believe that the rise in the *allegado* population is anything but a negative outcome of the way the housing market has been operating. Sharing is a sensible response to housing need, but it is not a response that should generally be encouraged.

Of course, it would be ideal if we could devise some kind of universal approach to the housing tenure question. In practice, however, there is no single rental housing strategy that is likely to be appropriate to the diverse conditions found in Caracas, Santiago and Mexico City, let alone to circumstances in the other cities of Latin America. Local conditions are simply too diverse; different housing recommendations are required for each city. Nevertheless, there is one broad lesson: governments should try to create as wide a range of housing alternatives as possible. In an environment of need a whole range of housing types can play a useful rôle, even some such as central city slums and lost cities, which have often been denounced as unsatisfactory. Without a satisfactory range of alternatives, some households will be forced into shelter entirely inappropriate to their needs.

The official encouragement that Latin American governments have given to self-help housing over the years has not been misguided. There is a continuing need for sites-and-services schemes and for squatter upgrading programmes. Such an approach, however, is no longer sufficient. Times have changed. The current economic recession means that large numbers of families find self-help development a much more difficult option. In addition, the sheer size of Latin America's largest cities means that many families find it difficult to live in the now distant urban periphery. We can no longer assume that every household can best satisfy its housing needs through home ownership. Current policies are therefore unbalanced. If subsidies were given to tenants as well as to owners, or alternatively if subsidies were no longer offered to owners, more families might want to rent accommodation. And, if tenants were to be assured of sufficiently long tenancies, the security seemingly offered by home ownership would appear in a less positive light. Similarly, if the rhetoric of government statements were less hostile to landlords and were matched by less glowing assertions of the advantages of owner-occupation, a better tenure balance might be achieved.

At present, few Latin American governments seek balance in their housing policies and most consistently favour a single-facetted housing solution. They encourage owner-occupation, sacrificing other forms of housing tenure on the altar of their favoured option. The effect is to narrow the range of housing alternatives, which leads inevitably to a decline in the living standards of the poor. In places, an increase in rental-housing accommodation would be an excellent way of broadening that range of alternatives.

References

Aaron, H. 1966. Rent controls and urban development: a case study of Mexico City. *Social and Economic Studies* **15**, 314–28.

Abu Lughod, J. 1976. The legitimacy of comparisons in comparative urban studies: a theoretical position and an application to North African cities. In *The city in comparative perspective*, J. Walton & L. H. Masotti (eds). New York: Halsted Press (John Wiley).

Aldunate, A., E. Morales, S. Rojas 1987. *Evaluación social de erradicaciones: resultados de una encuesta*. Materiales de Discusión. Santiago: FLACSO.

Amis, P. 1984. Squatters or tenants? The commercialization of unauthorised housing in Nairobi. *World Development* **12**, 87–96.

Amis, P. & P. Lloyd (eds) 1990. *Housing Africa's urban poor*. Manchester: Manchester University Press.

Angel, S. 1983. Land tenure for the urban poor. In *Land for housing the poor*, S. Angel et al. (eds), 110–40. Bangkok: Select Books.

Angel, S. & P. Amtapunth 1989. The low-cost rental housing market in Bangkok, 1987. *Habitat International* **13**, 173–85.

Azuela, A. 1987. De inquilinos a propietarios. Derecho y política en el Programa de Renovación Habitacional Popular. *Estudios Demográficos y Urbanos* **2**, 53–74.

Azuela, A. 1989. *La ciudad, la propiedad privada y el derecho*. Mexico City: El Colegio de México.

Azuela, A. 1990. *Institutional legal arrangements for the administration of land development in urban areas: the case of Mexico*. UNDP/UNCHS Urban Management Program, mimeo.

Bähr, J. & G. Mertins 1985. Desarrollo poblacional en el Gran Santiago entre 1970 y 1982. *Revista de Geografía Norte Grande* **12**, 11–26.

Banco Obrero 1973. *45 años del Banco Obrero, 1928–1973*, Caracas: Banco Obrero.

Barnes, S. 1987. *Patrons and power: creating a political community in Metropolitan Lagos*. Manchester: Manchester University Press.

Baróss, P. 1983. The articulation of land supply for popular settlements in Third World cities. In *Land for housing the poor*, S. Angel et al. (eds), 180–209. Bangkok: Select Books.

Baróss, P. & J. van der Linden 1990. *The transformation of land supply systems in third world cities*. Aldershot: Avebury.

BCV (Banco Central de Venezuela) 1978. *La economía venezolana en los últimos treinta y cinco años*. Caracas: BCV.

BCV 1981. *Anuario estadístico, 1980*. Caracas: BCV.

BCV 1984. *Anuario de séries estadísticas, 1983*. Caracas: BCV.

BCV 1986. *Informe económico 1985*. Caracas: BCV.

BCV 1987. *Informe económico 1986*. Caracas: BCV.

BCV 1988. *Informe económico 1987*. Caracas: BCV.

BCV 1990. *Informe económico 1989*. Caracas: BCV.

BCV 1991. *Informe económico 1990*. Caracas: BCV.

Beijaard, F. 1986. *On conventillos; rental housing in the centre of La Paz, Bolivia*. Free University of Amsterdam, Urban Research Working Paper, no. 5.

Bolívar, T. 1977. La producción de los barrios de ranchos y el papel de los pobladores y del estado en la dinámica de la estructura urbana del área metropolitana de Caracas. Paper presented at the Seminar on "Asentamientos Humanos Marginados", Jalapa, Mexico.

Bolívar, T. 1989. Los agentes sociales articulados a la producción de los barrios de ranchos. *Coloquio* 1, 143–62.

Bonduki, N. 1988. Crise de habitacão e a luta pela moradia no pos-guerra. In *As lutas sociais e a cidade: São Paulo: passado e presente*, L. Kowarick (ed.), 95–132. : São Paulo: Paz e Terra.

Bortz, J. 1984. Industrial wages in Mexico City 1939–75. Unpublished doctoral dissertation. University of California, Los Angeles.

Brown, J. 1972. *Patterns of intra-urban settlement in Mexico City: an examination of the Turner theory*. Cornell University Latin American Studies Program, Dissertation Series 40.

Camacho, O. O. & A. Terán 1991. *La propiedad y el inquilinato en cuatro barrios y casas de vecindad del área metropolitana de Caracas*. Caracas: Centro de Estudios Urbanos.

Carrera Damas, G. 1983. *Una nación llamada Venezuela*. Caracas: Monte Avila.

CED (Centro de Estudios del Desarrollo) 1990. *Santiago: dos ciudades*. Santiago: CED.

CENVI (Centro de la Vivienda y Estudios Urbanos) 1990. *Inquilinato y vivienda compartida en América Latina: investigación en cinco colonias populares de la Ciudad de México*. Mexico City: CENVI.

CEU (Centro de Estudios Urbanos) 1989. *Papel de trabajo: primeros avances de la investigación "El inquilinato y la vivienda compartida en ciudades latinoamericanas"*. Caracas: CEU, mimeo.

CEU (Centro de Estudios Urbanos) 1990. *La propiedad y el inquilinato en cuatro barrios y casas de vecindad del área metropolitana de Caracas, Informe Final*. Caracas: CEU.

CEU & OESE (Oficina de Estudios Socioeconómicos) 1977. *La intervención del estado y el problema de la vivienda*. Vol. I, *Introducción contexto nacional*. Caracas: CEU & OESE.

Chile, Ministerio de Vivienda y Urbanismo 1989. *Nueva ley de arriendo*. Ley Num. 18.101 D.0 31.178 de 29-01-1982. Santiago: Ed. Cumbres.

Cilento, A. 1989. 30 anos de financiamiento habitacional en Venezuela: cronología y crítica. *Coloquio* 1, 59–74.

Clark, W. A. V. & J. L. Onaka 1983. Life cycle and housing adjustment as explanations of residential mobility. *Urban Studies* 20, 47–57.

Cleaves, P. S. 1974. *Bureaucratic politics and administration in Chile*. Berkeley: University of California Press.

CNV (Concejo Nacional de la Vivienda) 1990. *Informe anual 1990*. Caracas: CNV.

Collier, D. 1976. *Squatters and oligarchs* Baltimore: Johns Hopkins University Press.

Collier, S. 1985. Chile. In *The Cambridge encyclopedia of Latin America and the Caribbean*, S. Collier, H. Blakemore, T. E. Skidmore (eds), 246–50. Cambridge: Cambridge University Press.

Connolly, P. 1977. *La producción de vivienda en la zona metropolitana de la Ciudad de México*. Mexico City: COPEVI.

Connolly, P. 1982. Uncontrolled settlements and self-build: what kind of solution? The Mexico City case. In *Self-help housing: a critique*, P. M. Ward (ed.), 141–74. London: Mansell.

Connolly, P. 1984. Las políticas hacía los asentamientos irregulares y la expansión metropolitana. In *Memoria del Encuentro para la Vivienda*, SEDUE & Gobierno del Estado de México (eds). Mexico City.

Connolly, P. 1987. La política habitacional después de los sismos. *Estudios Demográficos y Urbanos* 2, 101–20.

Connolly, P. 1988. Sector popular de vivienda: una crítica al concepto. *Medio Ambiente y Urbanización* 24, 3–14.

Conway, D. & J. Brown 1980. Intraurban relocation and structure: low-income migrants in Latin America and the Caribbean. *Latin American Research Review* 15, 95–125.

COPRE (Comisión Presidencial para la Reforma del Estado) 1987. *Lineamientos generales para una política de descentralización territorial en Venezuela*. Caracas: COPRE.

Cordera, R. & C. Tello (eds) 1984. *La desigualdad en México*. Mexico City: Siglo XXI.

Cordera Campos, R. & E. González Tiburcio 1991. Crisis and transition in the Mexican economy. In *Social responses to Mexico's economic crisis of the 1980s*, M. González de la Rocha & A. Escobar Latapí (eds), 19–56. San Diego: Center for US–Mexican Studies, University of California.

Cornelius, W. 1975. *Politics and the migrant poor in Mexico City*. Palo Alto, California: Stanford University Press.

Cornelius, W. & A. L. Craig 1988. *Politics in Mexico: an introduction and overview*. San Diego: Center for US–Mexican Studies, University of California.

Cornelius, W., J. Gentleman, P. H. Smith (eds) 1989. *Mexico's alternative political futures*. San Diego: Center for US–Mexican Studies, University of California.

Coulomb, R. 1985a. La vivienda de alquiler en las áreas de reciente urbanización. *Revista de Ciencias Sociales y Humanidades* VI, 43–70.

Coulomb, R. 1985b. *La legislación en materia de vivienda en arrendamiento para el distrito federal: situación actual (1985) y propuestas reglamentarias*. Cuadernos del CENVI. Mexico City: CENVI.

Coulomb, R. & C. Sánches Mejorada 1991. *¿Todos propietarios? Vivienda de alquiler y sectores populares en la Ciudad de México*. Mexico City: CENVI.

REFERENCES

Cuenya, B. 1986. El submercado de alquiler de piezas en Buenos Aires. *Medio Ambiente y Urbanización* **17** (Suplemento Especial), 3–8.
Cuenya, B. 1988. *Inquilinatos en la ciudad de Buenos Aires*. Buenos Aires: Centro de Estudios Urbanos y Regionales.

Daunton, M. J. 1987. *A property owning democracy? Housing in Britain*. London: Faber & Faber.
Delgado, J. 1988. El patrón de ocupación territorial de la ciudad de México al año 2000. In *Estructura territorial de la ciudad de México*, O. Terrazas (ed.), 101–41. Mexico City: DDF/Plazas y Valdés.
de Ramón, A. 1985. Vivienda. In *Santiago de Chile: características histórico ambientales, 1891–1924*, A. de Ramón & P. Gross (eds), 79–94. Monografías de Nueva Historia. London: Institute of Latin American Studies.
de Ramón, A. 1990. La población informal. Poblamiento de la periferia de Santiago de Chile, 1920–1970. *Revista EURE* **50**, 5–17.
Dietz, H. A. 1981. *Poverty and problem-solving under military rule: the urban poor in Lima, Peru*. Austin: University of Texas Press.
Doebele, W. 1987. The evolution of concepts of urban land tenure in developing countries. *Habitat International* **11**, 7–22.
Doling, J. 1976. The family life cycle and housing choice. *Urban Studies* **13**, 55–58.
Downs, A. 1983. *Rental housing in the 1980s*. Washington DC: The Brookings Institution.
Duhau, E. 1988a. Planeación metropolitana y política urbana municipal en la ciudad de México. *Estudios Demográficos y Urbanos* **3**, 115–42.
Duhau, E. 1988b. Política habitacional para los sectores populares en México. La experiencia de FONHAPO. *Medio Ambiente y Urbanización* **7**, 34–45.
Durand-Lasserve, A. 1986. *L'exclusion des pauvres dans les villes du tiers-monde*. Paris: L'Harmattan.

Eckstein, S. 1977. *The poverty of revolution: the state and the urban poor in Mexico*. Princeton, New Jersey: Princeton University Press.
Eckstein, S. 1990. Urbanization revisited: inner city slum of hope and squatter settlement of despair. *World Development* **18**, 165–82.
Edwards, M. 1981. Cities of tenants: renting as a housing alternative among the Colombian urban poor. Unpublished doctoral dissertation. University College London.
Edwards, M. 1982. Cities of tenants: renting among the urban poor in Latin America. In *Urbanization in contemporary Latin America*, A. G. Gilbert, J. E. Hardoy, R. Ramírez (eds), 129–58. New York: John Wiley.
Englander, D. 1983. *Landlord and tenant in urban Britain, 1834–1918*. Oxford: Oxford University Press.
España, L. P. & M. J. González 1990. Empobrecimiento y política social. *SIC* **522**, 62–4.
Ewell, J. 1984. *Venezuela: a century of change*. Palo Alto, California: Stanford University Press.

Figueroa, O. 1990. La desregulación del transporte colectivo en Santiago: balances de diez años. *Revista EURE* **49**, 23–32.

Fuad, K. 1974. Money isn't everything. *Business Venezuela* 32, 9.

FUNDACOMUN (Fundación para el Desarrollo de Comunidad y Fomento Municipal) 1978. *Inventario de los barrios pobres del area metropolitana de Caracas y el Departamento Vargas*. Caracas: FUNDACOMUN.

FUNDACOMUN 1985. *II inventario de los barrios en el Distrito Federal y el Estado Miranda, informe preliminar*. Caracas: FUNDACOMUN.

FUNDACOMUN 1988. *Proyecto sistema de análisis de información de barrios*. Caracas: FUNDACOMUN.

García, N. & M. López 1989. Vivienda obrera y gestión estatal (esquema histórico del Banco Obrero, 1928–1958). *Coloquio* 1, 39–58.

Garza, G. 1991. The metropolitan character of urbanization in Mexico, 1900–1988. Paper presented at the Institute of Latin American Studies, April 1991.

Garza, G. & M. Schteingart 1978. *La acción habitacional del Estado en México*. Mexico City: El Colegio de México.

Gazzoli, R., S. Agostinis, N. Jeifetz, R. Basaldúa 1989. *Inquilinatos y hoteles de Capital Federal y Dock Sur: establecimientos, población y condiciones de vida*. Buenos Aires: Centro Editor de América Latina.

Gil, E. 1988. La vivienda de interés social se construye en el interior. *Número* 397, 18.

Gil, E. 1991. Cifras millonarias se mueven en el mercado negro inmobiliaro. *El Diario de Caracas*, 30 August.

Gilbert, A. 1981. Pirates and invaders: land acquisition in urban Colombia and Venezuela. *World Development* 9, 657–78.

Gilbert, A. G. 1983. The tenants of self-help housing: choice and constraint in the housing markets of less developed countries. *Development and Change* 14, 449–77.

Gilbert, A. G. 1989. Housing during recession: illustrations from Latin America. *Housing Studies* 4, 155–66.

Gilbert, A. G. 1991. Comparative analysis: studying housing processes in Latin American cities. In *Housing the poor in the developing world: methods of analysis, case studies and policy*, A. G. Tipple & K. G. Willis (eds), 81–95. London: Routledge.

Gilbert, A. G. 1992. Third World cities: housing, infrastructure and services. *Urban Studies* 29, 435–60.

Gilbert, A. G. & P. Healey 1985. *The political economy of land: urban development in an oil economy*. Aldershot, England: Gower Press.

Gilbert, A. G. & A. Varley 1991. *Landlord and tenant: housing the poor in urban Mexico*. London: Routledge.

Gilbert, A. G. & P. M. Ward 1982. Residential movement among the poor: the constraints on Latin American urban mobility. *Transactions of the Institute of British Geographers* 7, 129–49.

Gilbert, A. G. & P. M. Ward 1985. *Housing, the state and the poor: policy and practice in three Latin American cities*. Cambridge: Cambridge University Press.

Gil Yepes, J. A. 1981. *The challenge of Venezuelan democracy*. New Brunswick, NJ: Transaction Books.

González de la Rocha, M. & A. Escobar Latapí (eds) 1991. *Social responses to Mexico's economic crisis of the 1980s*. San Diego: Center for US–Mexican Studies,

University of California.

Graham, C. 1991. *From emergency employment to social investment: alleviating poverty in Chile*. Washington DC: The Brookings Institution.

Green, G. 1988. The quest for tranquilidad: paths to home ownership in Santa Cruz, Bolivia. *Bulletin for Latin American Research* 7, 1–16.

Grimes, O. 1976. *Housing for low-income urban families: economics and policy in the developing world*. Baltimore: Johns Hopkins University Press.

Guerrero, M.-T. et al. 1974. *La tierra, especulación y fraude en el fraccionamiento de San Agustín*. Mexico City, mimeo.

Guzmán Cáceres, L. R. 1991. Políticas públicas y arrendamiento popular; el arrendamiento en Santiago de Chile, 1906–1950. *Revista EURE* 51, 59–62.

Handelman, H. 1979. *High-rises and shantytowns: housing the poor in Bogotá and Caracas*. American Universities Field Staff Report no. 9. Hanover, New Hampshire.

Haramoto, E. 1983. Políticas de vivienda social: experiencia chilena en las tres últimas décadas. In *Vivienda social: reflexiones y experiencias*, J. MacDonald (ed.), 75–152. Santiago: Corporación de Promoción Universitaria.

Harloe, M. 1985. *Private rented housing in the US and Europe*. London: Croom Helm.

Hellinger, D. C. 1991. *Venezuela: tarnished democracy*. Boulder, Colorado: Westview Press.

Hernández Laos, E. & M. Parás Fernández 1988. México en la primera década del siglo XXI: las necesidades sociales futuras. *Comercio Exterior* 38, 963–78.

Hoenderdos, W., P. van Lindert, O. Verkoren 1983. Residential mobility, occupational changes and self-help housing in Latin American cities: first impressions from a current research programme. *Tijdschrift voor Economische et Sociale Geografie* 74, 376–86.

Hoffman, M. L. et al. 1990. *Rental housing in Indonesia*. Washington DC: The Urban Institute.

Hure de Socorro, F. 1978. La dinámica de la ciudad. *Punto 60* 17, 85–94.

ICAL (Instituto de Ciencias Alejandro Lipschultz) 1988. Chile; una mentira cada ocho minutos. Transcript of a press conference given 10 December, 1987.

IDU (Institute de Desarrollo Urbano) 1990. *Investigación sobre vivienda compartida y en arriendo*. Santiago: IDU.

INAVI 1987. *Programa "Consolidación de barrios", lineamientos para el mejoramiento urbano y habitacional*. Caracas: INAVI

INC (Instituto Nacional del Consumidor) 1989. El gasto alimentario de la población de escasos recursos de la ciudad de México. *Comercio Exterior* 39, 52–8.

India (National Institute of Urban Affairs) 1989. *Rental housing in India: an overview*. NIUA Research Study Series, no. 31.

Iracheta, A. X. 1984. *El suelo, recurso extratégico para el desarrollo urbano*. Mexico City: UNAM.

Jaramillo, S. 1985. Entre el UPAC y la autoconstrucción: comentarios y sugerencias a la política de vivienda. *Controversia* 123–4.

Jones, G. A. 1991. The impact of government intervention upon land prices in

Latin American cities: the case of Puebla, Mexico. Doctoral dissertation. University of Cambridge.

Karst, K., M. Schwartz, A. Schwartz 1973. *The evolution of the law in the barrios of Caracas.* Berkeley: University of California Press.

Kemp, P. 1987. Some aspects of housing consumption in late nineteenth century England and Wales. *Housing Studies* **2**, 3–16.

Klak, T. 1992. Regional recession and working class shelter: Quito and Guayaquil during the 1980s. *Tijdschrift voor Economische en Sociale Geografie* **82**, in press.

Kowarick, L. (ed.) 1988. *As lutas sociais e a cidade: São Paulo: passado e presente.* São Paulo: Paz e Terra.

Kowarick, L. & C. Ant 1988. Cem anos de promiscuidade: o cortiço na cidade de São Paulo. In *As lutas sociais e a cidade: São Paulo: passado e presente*, L. Kowarick (ed.), 49–74. São Paulo: Paz e Terra.

Kusnetzoff, F. 1975. Housing policies or housing politics: an evaluation of the Chilean experieńce. *Journal of Interamerican Studies and World Affairs* **7**, 281–309.

Kusnetzoff, F. 1987. Urban and housing policies under Chile's military dictatorship 1973–1985. *Latin American Perspectives* **53**, 157–86.

Kusnetzoff, F. 1990. The state and housing in Chile – regime types and policy choices. In *Housing policy in developing countries*, G. Shidlo (ed.), 48–66. London: Routledge.

Lee-Smith, D. 1990. Squatter landlords in Nairobi: a case study of Korogocho. In *Housing Africa's urban poor*, P. Amis & P. Lloyd (eds), 175–88. Manchester: Manchester University Press.

Lemer, A. C. 1987. *The role of rental housing in developing countries: a need for balance.* World Bank Report, no. UDD-104. Washington DC: World Bank.

Lovera, A. 1987. La vivienda; los datos del problema. *SIC* **498**, 440–2.

Lowden, P. (no date). Villa El Cobre: a case study of Santiago's poor under military rule. MPhil. dissertation. Oxford University.

Lozano, E. E. 1975. Housing the urban poor in Chile: contrasting experiences under "Christian Democracy" and "Unidad Popular". *Latin American Urban Research* **5**, 177–96.

Malpezzi, S. & G. Ball 1991. *Rent control in developing countries.* World Bank Discussion Papers 129. Washington DC: World Bank.

Malpezzi, S. & S. Mayo 1987. User cost and housing tenure in developing countries. *Journal of Development Economics* **25**, 197–220.

Marcano, E. E. 1981. *Infraestructura de servicios de agua, cloacas y alcantarillados en el área metropolitana de Caracas.* Caracas: Publicaciones del Instituto de Urbanismo, Papel de Trabajo 16.

Marcano, E. E. 1987. El problema de los servicios. *SIC* **498**, 443–6.

Martínez Hernández, I. 1989. *Algunos efectos de la crísis en la distribución del ingreso en México.* Instituto de Investigaciones Económicas, Universidad Nacional Autonoma de México.

Martz, J. & D. Myers (eds) 1987. *Venezuela: the democratic experience*, 2nd edn. New York: Praeger.

Mayo, S. 1985. How much will households spend for shelter? *Urban Edge* **9**, 4–5.

Mayo, S. K., S. Malpezzi, D. J. Gross 1986. Shelter strategies for the urban poor in developing countries. *The World Bank Research Observer* 1, 183–203.

Mele, P. 1987. La dynamique de l'urbanisation de la ville de Puebla (Méxique) – de la ville a la région urbaine. Unpublished PhD thesis. Université de Paris III.

Meller, P. 1991. Adjustment and social costs in Chile during the 1980s. *World Development* 19, 1545–61.

Merrick, T. W. 1986. Population pressures in Latin America. *Population Bulletin* 41 (3).

Michelena, A. 1988. Pobreza y políticas sociales en Venezuela. Paper presented at the Forum on Estrategia para la Superación de la Pobreza.

Myers, D. 1978. Caracas: the politics of intensifying primacy. *Latin American Urban Research* 6, 227–58.

Necochea, A. 1987. El allegamiento de los sin tierra, estrategia de supervivencia en vivienda. *Revista Latinoamericana de Estudios Urbanos-Regionales (EURE)* 13–14, 85–100.

Necochea, A. & A. M. Icaza 1990. Una estrategia democrática de renovación urbana residencial: el caso de la comuna de Santiago. *Revista EURE* 48, 37–65.

Necochea, A. & P. Trivelli 1983. *Santiago Poniente: a case study of inner city blight and urban social dynamics.* UNCHS mimeo.

Negrón, M. 1982. Los orígines de la urbanización contemporanea en Venezuela: el crecimiento sin acumulación entre 1920 y 1945. *Urbana* 4.

Negrón, M. 1987. Cinquenta años de acción urbana en Caracas. Apuntes para un balance. *Vivienda* 7.

OCEI (Oficina Central de Estadística e Informática) 1986. *Situación habitacional en Venezuela.* Caracas: OCEI.

OCEI 1987. *Anuario estadístico de Venezuela, 1986.* Caracas: OCEI.

OCEI 1990. *Anuario estadístico de Venezuela, 1989.* Caracas: OCEI.

OCEI 1991. *Tiempo de resultados* 1 February. Caracas: OCEI.

Ogrodnik, E. 1984. Encuesta a los allegados en el Gran Santiago. *Revista de Economía* 22.

OMPU (Oficina Metropolitana de Planificación Urbana) 1972a. *Diagnóstico parcial y políticas generales en relación a los áreas de ranchos de Caracas, Informe prelimar.* Caracas: OMPU.

OMPU 1972b. *Plan General Urbano de Caracas 1970–1990.* Caracas: OMPU.

Ozo, A. O. 1990. The private rented housing sector and public policies in developing countries: the example of Nigeria. *Third World Planning Review* 12, 261–79.

Palacios, L. C., I. Niculescu, R. L. Clemente 1989. Vivienda y economía: una aproximación estructural. *Coloquio* 1, 75–106.

Payne, G. 1989. Informal housing and land subdivisions in third world cities: a review of the literature. Report prepared for the Overseas Development Administration. Oxford: Oxford Polytechnic.

Pérez Perdomo, R. & P. Nikken 1982. The law and ownership in the *barrios* of Caracas. In *Urbanization in contemporary Latin America*, A. G. Gilbert, J. E. Hardoy, R. Ramírez (eds), 205–30. New York: John Wiley.

Perló-Cohen, M. 1979. Política y vivienda en México 1910–1952. *Revista Mexicana de Sociología* **41**, 769–835.

Perló-Cohen, M. 1981. *Estado, vivienda y estructura urbana en el Cardenismo: el caso de la Ciudad de México*. Mexico City: Instituto de Investigaciones Sociales, Universidad Nacional Autónoma de México.

Perna, C. 1981. *Evolución de la geografía urbana de Caracas*. Caracas: Ediciones de la Facultad de Humanidades y Educación.

Pollack, M. & A. Uthoff 1989. Poverty and the labour market: Greater Santiago, 1969–85. In *Urban poverty and the labour market: access to jobs and incomes in Asian and Latin American cities*, G. Rodgers (ed.), 117–44. Geneva: International Labour Office.

Portes, A. 1989. Latin American urbanization during the years of the crisis. *Latin American Research Review* **25**, 7–44.

Portillo, A. J. 1984. *El arrendamiento de vivienda en la Ciudad de México*. Cuadernos Universitarios 5, Universidad Autónoma Metropolitana.

Potts, D. & C. C. Mutambirwa 1991. High-density housing in Harare: commodification and overcrowding. *Third World Planning Review* **13**, 1–25.

PREALC (Programa Regional del Empleo para América Latina y el Caribe) 1981. *Empleo y necesidades básicas: acceso a servicios urbanos y contratos públicos*. Santiago: PREALC.

Puente Lafoy, P. de la, E. Torres Rojas, P. Muñoz Salazar 1990. Satisfacción residencial en soluciones habitacionales de radicación y erradicación para sectores pobres de Santiago. *Revista EURE* **49**, 7–22.

Quintana, L. 1987. La vivienda popular en los 30 años de democracia. *SIC* **500**, 496–9.

Ragin, C. 1987. New directions in comparative research. In *Cross-national research in sociology*, M. L. Kohn (ed.), 57–76. Los Angeles: Sage.

Rakodi, C. 1987. Upgrading in Chawama, Lusaka: displacement or differentiation?. *Urban Studies* **25**, 297–318.

Ramírez, R., J. Fiori, H. Harms, K. Mathey 1991. *The commodification of self-help housing and state intervention. Housing experiences in the "barrios" of Caracas*. UCL Development Planning Unit Working Paper, no. 26. London: DPU, University College London.

Ray, T. 1969. *The politics of the barrios of Caracas*. Berkeley: University of California Press.

Riofrío Benavides, G. 1978. *Se busca terreno para proxima barriada: espacios disponibles en Lima 1940, 1978, 1990*. Lima: DESCO.

Rodríguez, A. 1983. Como gobernar ciudades o principados que se regían por sus propias leyes antes de ser ocupados. *Revista de la Sociedad Interamericana de Planificación* **17**, 135–55.

Rodríguez, A. 1989. Santiago, viejos y nuevos temas. In *La investigación urbana en América Latina: caminos recorridos y por recorrer - Estudios nationales*, F. Carrión (ed.), 203–36. Quito: Ciudad.

Rodríguez, J. I. 1976. The savings and loan housing market in Venezuela during the 'sixties and early 'seventies. Unpublished doctoral dissertation. Vanderbilt University.

Rojas Gutiérrez, C. 1991. Avances del Programa Nacional de Solidaridad. *Comercio Exterior* **41**, 443–6.

Sachs, C. 1990. *São Paulo: politiques publiques et habitat populaire*. Paris: Editions de la Maison des Sciences de l'Homme.

Saunders, P. 1990. *A nation of home owners*. London: Unwin Hyman.

Schteingart, M. 1990. *La producción del espacio habitable; estado, empresa y sociedad en la ciudad de México*. Mexico City: El Colegio de México.

SHCP (Secretaría de Hacienda y Crédito Público) 1964. *Programa financiero de vivienda*. Mexico City: SHCP.

Silva Michelena, J. A. (ed.) 1987. *Venezuela hacía el 2000: desafíos y opciones*. Caracas: Editorial Nueva Sociedad.

Soto, J. 1987. *El acceso a la vivienda de los sectores más pobres de la región metropolitana*. Universidad Católica, mimeo.

Stann, J. 1975. Transportation and urbanization in Caracas, 1891–1936. *Journal of Interamerican Studies and World Affairs* **17**, 82–100.

Stewart, W. S. 1987. Public administration. In *Venezuela: the democratic experience*, 2nd edn, J. Martz & D. Myers (eds), 218–41. New York: Praeger.

Stolarksy, N. 1982. *La vivienda en el Distrito Federal*. Mexico City: DDF, Dirección General de Planificación.

Stren, R. E. 1990. Urban housing in Africa: the changing rôle of government policy. In *Housing Africa's urban poor*, P. Amis & P. Lloyd (eds), 35–54. Manchester: Manchester University Press.

Sudra, T. 1976. Low-income housing system in Mexico City. Unpublished doctoral dissertation. Massachusetts Institute of Technology.

Tipple, A. G. 1988. *The development of housing policy in Kumasi, Ghana, 1901 to 1981*. University of Newcastle-upon-Tyne, Centre for Architectural Research and Development.

Tipple, A. G. & K. G. Willis 1991. Tenure choice in a West African city. *Third World Planning Review* **13**, 27–46.

Trivelli, P. 1986. Access to land by the urban poor: an overview of the Latin American experience. *Land Use Policy* **3**, 101–21.

Trivelli, P. 1987. Intra-urban socio-economic settlement patterns, public intervention, and the determination of the spatial structure of the urban land market in Greater Santiago, Chile. Unpublished doctoral dissertation. Cornell University.

Turner, J. F. C. 1967. Barriers and channels for housing development in modernizing countries. *Journal of the American Institute of Planners* **33**, 167–81.

Turner, J. F. C. 1968. Housing priorities, settlement patterns, and urban developing in modernizing countries. *Journal of the American Institute of Planners* **34**, 354–63.

United Nations 1979. *Commission on human settlements*. Document HS/C/2/3.

UN, Department of International Economic and Social Affairs 1988. *Housing and economic adjustment*, New York: United Nations.

UNCHS 1989. *Strategies for low-income shelter and services development: the rental housing option*. Nairobi: UNCHS.

REFERENCES

UNECLAC (United Nations Economic Commission for Latin America and the Caribbean) 1984. *Preliminary overview of the Latin American economy 1984.* Santiago: UNECLAC.

UNECLAC 1988. *Statistical yearbook for Latin America and the Caribbean, 1987.* Santiago: UNECLAC.

UNECLAC 1989. *Preliminary overview of the Latin American economy 1989.* Notas sobre la economía y el desarrollo 485/6. Santiago: UNECLAC.

UNECLAC 1990. *Preliminary overview of the Latin American economy 1990.* Santiago: UNECLAC.

van Lindert, P. 1991. Moving up or staying down? Migrant–native differential mobility in La Paz. *Urban Studies* 28, 433–63.

van Lindert, P. & A. van Western 1991. Household shelter strategies in comparative perspective: evidence from low-income groups in Bamako and La Paz. *World Development* 19, 1007–28.

Varley, A. 1985. Ya somos dueños: ejido land development and regularisation in Mexico City. Unpublished doctoral dissertation. University College London.

Walton, J. 1989. Debt, protest, and the state in Latin America. In *Power and popular protest: Latin American social movements,* S. Eckstein (ed.), 299–328. Berkeley: University of California Press.

Ward, P. M. 1981. Political pressure for urban services: the response of two Mexico City administrations. *Development and Change* 12, 379–407.

Ward, P. M. (ed.) 1982. *Self-help housing: a critique.* London: Mansell.

Ward, P. M. 1990a. *Mexico City.* London: Belhaven.

Ward, P. M. 1990b. The politics of housing production in Mexico. In *The international handbook of housing policies and practices,* W. van Vliet (ed.), 407–36. Boulder, Colorado: Greenwood Press.

Ward, P. M., E. Jiménez, G. Jones 1991. Residential land price changes in Mexican cities and the affordability of land for low-income groups. Paper presented at Fitzwilliam College, Cambridge, July.

Wilkie, J. W., D. E. Lorey, E. Ochoa (eds) 1988. *Statistical abstract of Latin America,* **26.** Los Angeles: UCLA Latin America Center Publications.

Willis, K. G., S. Malpezzi, A. G. Tipple 1990. An econometric and cultural analysis of rent control in Kumasi, Ghana. *Urban Studies* 27, 241–58.

World Bank 1989. *World development report 1989.* Washington DC: World Bank.

World Bank 1990. *World development report 1990.* Washington DC: World Bank.

Yujnovsky, O. 1984. *Claves políticas del problema habitacional Argentino.* Buenos Aires: Grupo Editor Latinoamericano.

Index

family structure and tenure 6, 41, 42,
44–5, 48, 52, 83–4, 86–7, 91, 92, 93,
126, 129–30, 134, 135, 138, 142–3,
144, 148, 155
FOGA 31
FONHAPO 33, 37
FOVI 30, 31
FOVIMI 32
FOVISSSTE 32
Frei Montalva, Eduardo 57, 58, 66,
69–70, 74, 76, 77
Fresno, Juan Francisco 76
FUNDACOMUN 117, 119, 120–1

Gómez, Juan Vicente 98
Guadalajara 13, 47, 143, 147
Herrera Campíns, Luis 98
housing,
conditions and tenure 44–5, 87,
130–1, 144
public housing 2, 9, 11, 24, 28,
32–4, 60–1, 64, 68–74, 82–9, 99,
102, 103, 108, 114, 116–8, 140, 153
subsidies 5, 11, 13, 28, 30, 33, 60,
66, 67, 68–74, 82, 88–9, 108, 114,
115–6, 142, 153, 158–9, 160
tenure trends 1, 25–6, 29, 31, 56,
62–4, 107–8, 139–40

INAVI 117, 118
Ibáñez del Campo, Carlos 58, 68
income and tenure (see also
rent:income ratios) 41, 42–3, 49,52,
56, 81, 83, 84–5, 91, 92, 93, 95,
126–8, 135, 138, 141–2, 144
income distribution 17–19, 58–61, 93,
96–9, 153, 158
INDECO 30, 32
inflation 8–9, 10, 18, 29, 57, 59, 65,
98, 100, 115, 116, 151
INFONAVIT 30, 32
IVSS 113, 115
infrastructure, see services

La Paz 146, 147
La Victoria 68
land
ejidos 20–3, 34–7, 38–9, 79, 123

illegal subdivisions 5, 10–11, 13,
20–2, 22–4, 34–6, 38, 51, 61, 79,
124, 140, 142, 145, 149, 154
invasions 5, 10–11, 13, 22, 34–7,
39–40, 51, 58, 60–1, 68, 69–71,
74–7, 79, 82, 83, 95, 103, 118–23,
124–5, 140, 142, 145, 149, 153–4
legalization of tenure 11, 20, 35, 40,
53, 77, 83, 121–3, 159
prices 2, 3, 28, 31, 33, 47, 77, 96,
99, 117, 153–4
landlords 1–2, 4–5, 6, 40, 52–55, 56,
64–5, 67, 92–5, 109–14,134–7, 138,
150–3, 157–9, 160–1
landlord–tenant relations 1–2, 6,
54–5, 94–5, 136–7, 151–2, 157
law and rental housing 2, 6, 29–31,
54–5, 64–5, 66–7, 94–5, 107–8,
109–14, 120, 136–7, 151, 157–8
legalization, see land
Leoni, Raúl 98
Lima 153
López Contreras, Eleazar 98
López Portillo, José 36
Lusinchi, Jaime 98, 122

Maximilian 27
Medina Angarita, Isais 98, 109
methodology 11–16
Metro 9–10, 20, 24, 40–1, 107
Mexico City 3, 4, 5, 8–16, 17–56, 61,
86, 89, 95, 99, 109, 139–61
migrants and tenure 41, 47–8, 49, 64,
90, 126, 131, 132–3, 134, 137–8,
143–4, 146–7, 148, 149, 151–2, 153
migration 9, 10, 61, 64, 118, 132, 149,
153
MINVU 74, 82

Nairobi 4
Netzahualcóyotl 22, 24, 35, 40

oil 10, 17–19, 96–100, 103, 104, 108,
115
Operación Sitio 69–71, 73, 74, 82–3
owner-occupation,
household preference for 1–2, 45–8,
71, 88–90, 131–3, 137–8, 140,